The Quest for Ascendant Quality

An Introduction to
Contemporary Philosophy and Methods
for Strategically Orchestrating the Transformation
Towards & Beyond Quality Excellence
in Everything you do

Victor Elias

An

On Queue Universal Educational Services & Training

Publication

Elias, Victor. *The Quest for Ascendant Quality: An Introduction to Contemporary Philosophy and Methods for Strategically Orchestrating the Transformation Towards & Beyond Quality Excellence in Everything you do.* Sparta, NJ: On QUEST, 2014.

Library of Congress Control Number: 2015902273

ISBN: 978-0-9990801-1-5

The following trademarks of Victor Elias are used in this book:
> The Quest for Ascendant Quality™
> Themes of Performance™
> Thematic Risk Assessment™
> The Balanced Enterprise Strategic Themes™ Framework
> The BEST Framework™
> Saile Charting™

The following service mark of Victor Elias is used in this book:
> Process Enrichment®

Process Enrichment is registered in the U.S. Patent and Trademark Office.

To find out more about Process Enrichment Boot Camp seminars see:
linkedin.com/in/processenrichmentbootcamp
https://www.facebook.com/Process-Enrichment-Boot-Camp-394144370770329/

Please direct inquiries to the author's attention care of On QUEST.

Victor Elias, c/o
On QUEST
677 West Mountain Road
Sparta, NJ 07871

E-mail: OnQUEST@Outlook.com

To Lisa, Victoria, and Oliver

and to all of us,
who journey together,
on the quest for ascendant quality

Contents

1 – The Quest for Ascendant Quality...1

2 – Elements of Strategic Communication & Quality............................45

Case Studies in Quality's Six Themes of Performance
▶ The U.S.S. Indianapolis
▶ Aluminized Explosives ("Most Secret")
▶ The LZ 129 Hindenburg
▶ The Battle of the Atlantic
▶ The Final Voyage of the Titanic
▶ The Battle of Britain

3 – The Attraction of Creative Vision Towards & Beyond
Quality Excellence...129

Case Study:
▶ Operation Focus: The Air Battle at the Opening of the 1967
Six Day War..163

4 – The Nature of Risk...177

Exercise:
▶ Composing a Risk Management Strategy to Circumvent and Counteract
the Risk of Improvised Explosive Devices Emplaced in Roadways

Case Study:
▶ The Parable of the Fearless Slugger

Case Study:
▶ NASA's Space Shuttle Program and The Fearlessness Effect

5 – The Ascendant Theory of Motivation 235

6 – Reflections & Vision: The Quest for Ascendant Quality.............265

Preface

Welcome to the quest for ascendant quality. I hope you will enjoy the adventure of reading this book as much as I enjoyed the adventure of writing it.

I have often found that when I set out to write something, I learn more about the topic after my pen hits the paper than I knew before starting to write – and my adventure in writing this book has been no exception. Although, due to my lifelong passion for the art and science of quality, I started out with insights that were new and enlightening to the several close friends and associates I shared them with, I found that the processes of developing a narrative, indulging my intellectual curiosity, learning, and writing, gave rise to a completely different way to understand quality than I knew before. It was necessary to develop this understanding because, I realized, even though quality is foremost in the thoughts of everyone involved in creative effort this common term has never been adequately defined in a correct and useful way so that an understanding of quality could improve the ability to attain performance that meets customer needs and advances mission success. Through methodical research, the best of which is presented in this book, I developed the unique theory that the concepts of strategy, quality, risk, and the aspiration to improve are elements of an interrelated system. As the continually supportive results of my research led me to a steadfast belief in my theory, I experienced a kind of stress that, I imagine, is the good kind. It's the kind of stress that arises when you know that you're onto a set of ideas that can serve to enable anyone with a creative vision to improve their ability to create or acquire products and services which will be capable of performing with quality excellence in the service of their strategy. Through a willingness to follow paths of research which, at first, seemed eccentric – a completely new conceptual system emerged that is logical, is applicable to any creative effort whatsoever, and, as I hope you will agree, is consistent with what most of us actually believe the relationships defining quality to be. One realization that I encountered, quite unexpectedly, as I tried to understand people's aspiration to improve, is that the need to be on the quest for ascendant quality is a fundamental human need – unique in its ability to orchestrate and refocus all other needs, unique in its irrepressibility and consistency as a lifelong influence on behavior, and unique in its ability to create common purpose and alliances among people. It is this fundamental need which drives the aspiration to improve strategy and quality and to reduce risk as part of a complete, humanistic, model of quality's conceptual system. In a very practical sense, we are likely to discover a great many ways that an improved understanding of this conceptual system can, over the long term, serve to reliably improve our ability to improve our civilization and our quality of life.

The Quest for Ascendant Quality is presented in six chapters that build progressively in providing insight to the historical significance, fundamental purpose, and meaning of quality and to the manner with which this knowledge can be used effectively to improve creative effort. For this reason, I recommend reading through each chapter sequentially.

In chapter 1, The Quest for Ascendant Quality, we discuss the influence of commerce, historical events and circumstances, and innovation which served to improve our quality of life and shape our concept of quality. We introduce the Process Enrichment® philosophy as a basis for a new way of thinking about quality in our creative effort and in our collaborative relationships. We introduce the risks of poor performance inherent in many of the last century's conceptual ideas about strategy, quality, and risk as a justification for the introduction of the Process Enrichment philosophy's new concepts and methods for quality improvement.

In chapter 2, Elements of Strategic Communication and Quality, we explore the ways in which strategic communication acts to advance organizational motives and intentions, improve quality, productivity, and value for customers. We study the architecture of successful strategic communication, we introduce new definitions of quality and strategy, and we are introduced to new concepts that enable strategic communication to perform successfully – including quality's six Themes of Performance and, also, the Balanced Enterprise Strategic Themes (BEST) framework. Six prototypical case studies build depth of understanding of each of quality's six Themes of Performance and present unique perspectives and new insight to the impact of different aspects of quality on strategy as revealed by significant events in world history. Each case study offers lessons and implications for contemporary organizations. A Thematic system for identifying performance measurements is also introduced.

In chapter 3, The Attraction of Creative Vision Towards & Beyond Quality Excellence, we introduce and discuss a new definition of Quality Excellence and the manner with which it can be used as a standard of performance. We introduce Saile Charting as a means of visually communicating the status of quality and risk in relationship to strategic performance targets. We discuss how to identify risks of poor performance and how to influence the effort to implement necessary improvements. We use Saile Charts to prioritize improvement effort based on the market impact of implementing necessary improvements. We use Saile Charts to assess and compare performance quality relative to customer strategy and relative to the performance quality of competing products or services. We use Saile charts to evaluate competitive position and to identify and prioritize performance improvements that improve the market value of a product and service. We define ascendant quality and discuss the quest for ascendant quality as a driver of innovation in creative effort.

In chapter 4, The Nature of Risk, we introduce a new definition of risk and present a method for evaluating the impact of risk on quality and strategy over a risks' life cycle. We discuss how risk can be used to drive implementation of necessary quality improvements. We also discuss how reducing risk leads to improved competitive position. As an exercise, we compose a risk management strategy to circumvent the notorious Improvised Explosive Devices (IEDs) as they are used to sabotage roadway travel in Iraq and Afghanistan. We introduce the Fearlessness Effect and the proposition that the Fearlessness Effect was the cause of the destruction of NASA's Space Shuttles Challenger in 1986 and Columbia in 2003.

In chapter 5, The Ascendant Theory of Motivation, we introduce new insight to the work of Maslow and Herzberg and the concepts of satisfaction, dissatisfaction, and motivation. We introduce the Ascendant Theory of Motivation and discuss how this theory improves our understanding of the fundamental and basic human needs. We introduce the new concepts of personal strategy and Coascendancy. We introduce the Coascendancy Totem to structure our discussion on the influences that cause people to create high quality products and perform high quality services with personal gratification. We also discuss the obstacles that should be overcome in order for organizations and their employees and customers to align their behavior with organizational strategy that assures high productivity and customer value.

In chapter 6, Reflections and Vision: The Quest for Ascendant Quality, we reflect on the historic impact of the quest for ascendant quality and present a vision of the potential impact of our new concepts and principles on creative effort now and in the future.

Whether your intent in reading this book is to improve the quality of creative effort in your personal life or in your business life, the philosophy and methodology presented in this book is intended to be useful to you for the rest of your life – wherever your quest for ascendant quality takes you.

I hope you will enjoy your adventure in reading this book and in your quest for ascendant quality.

The Quest for Ascendant Quality

Chapter 1:

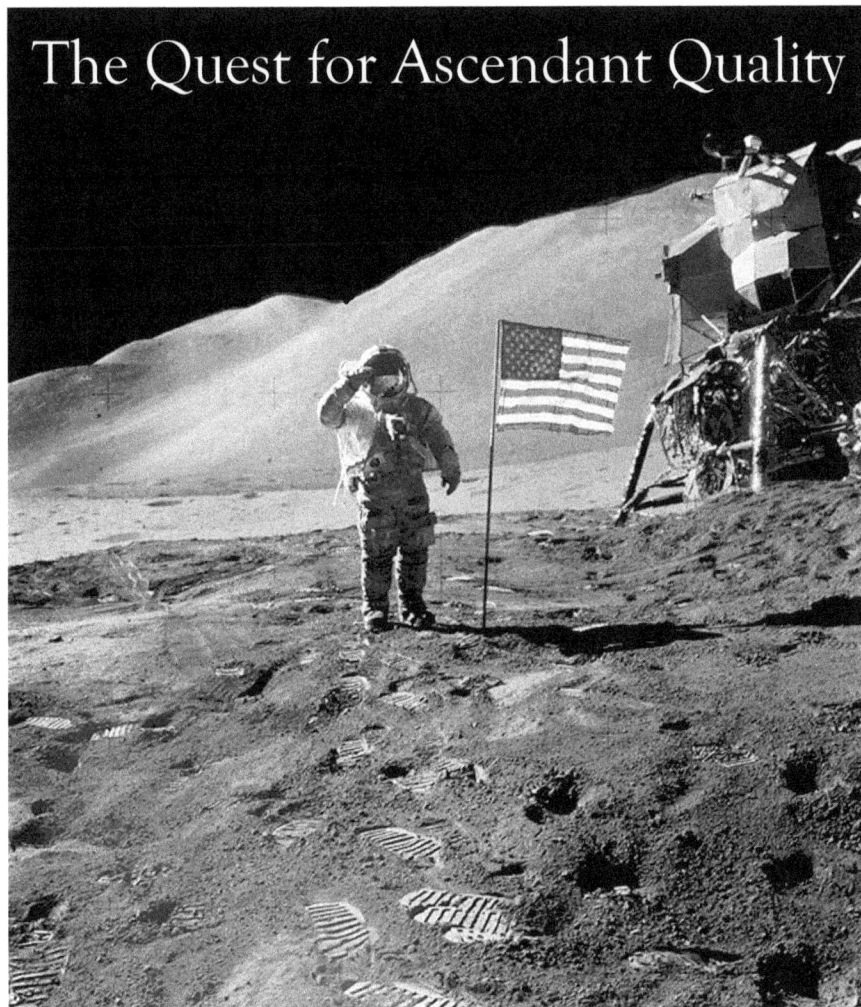

The Quest for Ascendant Quality

Wherever life is found, life is found on the quest for ascendant quality. For the simplest to the most complex of living beings, life itself compels the ceaseless, multi-generational quest for ascendancy – to advance the quality of life in the world in a beneficial direction and in so doing achieve gratification that makes time and life enjoyable and purposeful.

Chapter 1:

The Quest for Ascendant Quality

- A brief history of societal progress driven by the Quest for Ascendant Quality

- Retrospective on the strengths and weaknesses of the last century's concepts of strategy, quality and risk

- Introducing the Process Enrichment philosophy's concepts of:

    Strategy ◆ Risk
    Quality ◆ Coascendancy

- The significance of the art and science of quality improvement to commerce in and beyond this century

The quest for ascendant quality is a timeless story of the inalienable and irrepressible spirit of life. From the first microscopic plants and animals of eons past to our contemporaries, who scan the universe for the next green pasture, we share the enduring need to survive and the aspiration to thrive. Our historic pursuit of better ways to meet our individual and collective needs has served to enable civilizations to form, to learn from each other, and to transform over time to produce the high quality of life that we now enjoy. The evidence of this ever progressing transformation is found in artifacts deep in the history of civilization. While human conflict, natural disaster, disease, and our lifespan may intervene, the quest for ascendant quality carries on and is carried forward by successive generations as the spirit of life fills our world with the creations most desired by all.

It's through the prism of this long view of the arc of history that we reveal, in splendid form, the uniquely human story of progress driven by the quest for ascendant quality. From a human standpoint we may observe that living on the quest for ascendant quality is an endeavor that our minds and bodies are well suited for. It's because of our abilities to sense and discern various forms and grades of quality, in relation to our needs, that we can create or acquire quality excellence in the goods and services that we desire. Our five senses help us to hone-in on quality excellence so that we know a fine silk when we touch it; we know a fine meal when we taste it; we know a fine painting when we see it; we know a fine rose when we smell it; and we know a fine orchestra when we hear it. Taken together with our capabilities to communicate, and to empathically sense and understand quality excellence from each other's viewpoints, we are finely attuned in the natural capabilities of mind and body that enable us to envision and enact many forms of mutually beneficial creative effort. When we skillfully use our many natural capabilities as we acquire products and services that best meet our needs or as we produce the products and services that customers are ready, willing, and able to purchase, we achieve quality excellence, commerce thrives, communities flourish, and we achieve harmony with our quest for ascendant quality.

The civilized tradition of the exchange of products and services among people plays a key role in the quest for ascendant quality. It's through the exchange of products and services that we help each other improve our quality of life. Each party to a fair, ethical, exchange gains the benefit of different forms of value based on the talents, efforts, and resources of their trading partner. In this way, each side of an exchange can enjoy products and services that sustain or improve their quality of life that they would have difficulty creating on their own. The benefits of commerce on an international scale were expressed nearly 200 years ago by the renowned economist David Ricardo.

> "Under a system of perfectly free commerce, each country naturally devotes its capital and labour to such employments as are most beneficial to each. This pursuit of individual advantage is admirably connected with the universal good of the whole. By stimulating industry, by regarding ingenuity, and by using most efficaciously the peculiar powers bestowed by nature, it distributes labour most effectively and most economically: while, by increasing the general mass of productions, it diffuses general benefit, and binds together by one common tie of interest and intercourse, the universal society of nations throughout the civilized world."[1]
>
> David Ricardo
> 1817

While the potential for improvements in our quality of life is as boundless as time and creativity, the development and propagation of improvements throughout our world depends greatly on the effectual functioning of commerce. Because of the centrality of commerce to the advancement of our quality of life, there is a great cultural desire to establish high quality conditions of trade which keep trade fair, ethical, and efficient and which minimize risks of poor performance for both customers and suppliers in their trading activities. While legal and ethical standards in trade are essential, the best assurance that the risk of poor performance in commerce is at a minimum comes from the talent of the trading parties themselves in their ability to understand the elements of quality. When the customers in a market can properly assess quality they can make informed decisions that serve their best interests and that also advance the prosperity and ability to succeed of the suppliers that produce the products or services that most closely achieve their concept of quality excellence. In this way,

---

[1] Ricardo, David. *On the Principles of Political Economy, and Taxation*, London: John Murray, 1817.

over time, the ability to recognize and produce quality excellence raises community expectations, and raises capabilities to achieve quality excellence, and quality continues to ascend. On the other hand, when the customers in a market cannot appreciate the benefit of a new or improved product or service or, when a supplier cannot recognize that improvements in their products or services are needed by customers, progress stalls and there is, in consequence, a loss to society of the improvement in quality of life while this condition persists.

The impact of this condition is reminiscent of a scene in science fiction from the original Star Trek television series. In this scene, captain Kirk and his first officer Mr. Spock beam down from the starship Enterprise, which is orbiting the planet Organia, to the planet surface where they materialize in a quaint medieval style village. After Spock makes an assessment of the planet using his tricorder the following dialog occurs:

> "SPOCK: Captain, our information on these people and their culture was not correct. This is not a primitive society making progress toward mechanization. They are totally stagnant. There is no evidence of any progress as far back as my tricorder can register.
> KIRK: That doesn't seem likely.
> SPOCK: Nevertheless, it is true. For tens of thousands of years, there has been absolutely no advancement, no significant change in their physical environment. This is a laboratory specimen of an arrested culture.
> KIRK: Thank you, Mister Spock. That might be useful."

Fascinating; but, of course, in good science fiction, life in an "arrested culture" cannot truly be as it seems because that would be contrary to the philosophical premise that this chapter began with: "Wherever life is found, life is found on the quest for ascendant quality" and it sounds as though Captain Kirk read this chapter. The truth that it was an illusion that Organia was an arrested culture was ultimately revealed by Ayelborne, the chairman of Organia's Council of Elders:

"AYELBORNE: Millions of years ago, Captain, we were humanoid like yourselves, but we have developed beyond the need of physical bodies. That of us which you see is mere appearance for your sake."[2]

Star Trek, Errand of Mercy
3/23/1967

Outside of science fiction, cultures cannot be arrested in progress for "tens of thousands of years" because, ultimately, the quest for ascendant quality drives improvement – sometimes accelerated by the great force of competition – and valued improvements are efficiently distributed, where needed, through the effectual functioning of commerce.

In a healthy market, the story of a business whose journey takes it from good to great may end for that business at great because each great business's best performance eventually becomes merely acceptable as customers increasingly demand better and, in time, a new concept of great emerges. Add to the customers upgraded concept of great the potential for attraction into the market of increased competition, drawn by the once great company's success, and the once great business must either fade into obsolescence or reclaim its greatness by revitalizing its understanding of quality and its passion to improve and create quality excellence once again.

While businesses are normally the creative instruments of ascendant quality in commerce, when they fail to sustain an acceptable level of quality, as the threshold for acceptable quality in a culture grows, they can sometimes have the opposite effect. For example, when the once great and now obsolescent companies are not adequately challenged by competition they bear the potential to exploit their position of power in the market by increasing prices, by inhibiting the growth of quality in the products and services they sell, or by intentionally reducing rather than improving quality – sometimes with a negative impact on safety – in order to reduce costs and improve profit. In the interest of survival or continued profitability the once great companies may resort to political and social intrigues that are contrary to our nations will and well being. When a business intentionally thrives on such a "pocket of arrested culture," at the

---

[2] *"Errand of Mercy."* Star Trek. Season 1, episode 27, NBC. 3/23/1967.

expense of an ascendant culture, we can see that their advertisements extolling their virtues, their concern for quality, safety, and the environment, and their aspirations to improve, are all, to use Ayelborne's expression, "mere appearance" for the sake of customers.

The venerable maxim "As GM goes, so goes America" makes the story of General Motors (GM) Corporation, which rose to greatness as the worlds largest automobile manufacturer, on the enthusiasm of satisfied customers, only to succumb to obsolescence and customer abandonment, a momentous and compelling lesson to us all of the risk and consequences our nation faces when a business transforms to exploit its market power and loses track of its historic pursuit of ascendant quality.

GM was propelled to greatness in the 1920's and 1930's when GM President Alfred Sloan's market segmentation strategy, with an intention to produce "a car for every purse and purpose," succeeded, as no other automaker had before, in offering customers an opportunity to purchase a car that was appropriate to their financial resources and enabling of their purposes, or strategies, for use of the car – commuting, shopping, vacationing, and travelling in style.

> "General Motors was becoming large through a process of evolution, but only because it was rendering a service to community. As its volume of business expanded it became able to do more for workers, stockholders and customers."[3]
>
> Alfred Sloan

By understanding the purposes for which customers used their cars and through innovation and quality excellence in design, GM brought a greater awareness of the sensations of style and distinction to the automobile. These were qualities which the competing Ford Motor Company would not match due to their strategy of producing only one, low cost, car model in only one color – black. It was only natural for people to desire more and better styling and utility

---

[3] Sloan, Alfred P. Jr. and Boyden Sparks, *Adventures of a White-Collar Man*, (New York: Doubleday, Doran & Co., Inc.), 1941), p. 144.

in their cars when they could afford it. This was especially true for car buyers on their second or third car whose interests and appreciation of a broader concept of quality was maturing with their growing exposure to a variety of interesting new cars. As Mr. Sloan wrote of this period in his autobiography:

> "Middle-income buyers, assisted by the trade-in and installment financing, created the demand, not for basic transportation, but for progress in new cars, for comfort, convenience, power, and style."[4]
>
> Alfred Sloan

Fords' loss of market leadership to GM is typical of the consequences of the risk all companies take, in a competitive environment, by not offering each form of quality which customers demand at the level of satisfaction demanded. The availability of GM's innovative cars, which offered "comfort, convenience, power and style," and the evolution of a new concept among customers of what a great car is, created a market of car buyers which Ford's product strategy couldn't serve. GM's growing market share eventually forced Ford to transform from their poorly performing product strategy to a more defensible strategy that better represented the new concept of a great car, which customers favored, but it took them a long time. These were the times of growing market power for GM as they surpassed Ford to become the world's largest automobile producer.

A significant yet subtle breach in the trust between GM and its customers emerged during the 1950's as efforts to increase sales and profit became more important than creating quality excellence. This change in GM's focus was described, sketchily at first, by Harley Earl, GM's Vice President of Design, who reportedly introduced the strategy that came to be known as "planned obsolescence" to his firm.

> "Design these days means taking a bigger step every year. Our big job is to hasten obsolescence. In 1934 the average car ownership

---

[4] Sloan, Alfred P. Jr. *My Years with General Motors* (New York: Doubleday and Co., Inc., 1964), p. 163.

span was 5 years; now it is 2 years. When it is a year, we will have a perfect score."[5]

Harley Earl
1955

GM's 12th President, John Gordon, was less sketchy than Mr. Earl in defending the strategy of planned obsolescence. During a luncheon, as reported by The Wall Street Journal, he spoke revealingly about the intended consequences of planned obsolescence as he defended the costs of annual model changes, which were the main initiative in support of planned obsolescence.

"He argued that if manufacturers didn't change their models each year, the 20 million new car buyers who come into the market every three years might buy a new car only once in six years. "In other words they would continue to use the same car as long as it gave satisfactory service. There would be no incentive to replace it. Obviously, this would cut our annual sales in half," he said. "With volume reduced, each car would cost more to produce, and I am sure this increased cost would more than offset any saving resulting from not changing models."[6]

John Gordon
1960

At the very time when cosmonauts and astronauts began orbiting the earth, in spacecraft engineered with the flawless reliability and precision of a fine watch, GM openly regretted that their cars gave customers satisfactory service for too long. For GM, planned obsolescence meant the substitution of attractive and enjoyable stylistic features, each year, for other forms of perhaps more necessary performance quality – such as reliability.

Planned Obsolescence worked because changes in styling helped to sell more cars more often as customers' pride made them want to "keep up with the Joneses" and also because quality excellence in stylistic design creates a

---

[5] Mitarachi, Jane Fiske. "Harley Earl and His Product: The Styling Section." Industrial Design 2 (October 1955): p. 52

[6] The Wall Street Journal, "GM to Raise '61 Capital Spending to $1,250,000,000," Nov. 1, 1960

psychological impression, or a façade, of broader high quality. Another reason for the success of planned obsolescence is that as lower reliability in their cars became evident to customers, they were compelled to buy again, more frequently, to avoid getting stuck with a car that tended to need mechanical work and parts replaced and that rapidly decreased in resale value. These conditions made the period of pride in ownership ever more fleeting as relatively new cars showed their age compared to the latest and greatest model and became dissatisfying to own since they no longer conveyed status. Products with an unexpectedly high risk of poor performance and with rapid onset of customer dissatisfaction, reveal the classic symptoms characteristic of planned obsolescence.

The success, or at least the profitability, of planned obsolescence significantly accelerated GM's shift away from quality excellence. Instead of continuing a customer focused, ascendant, approach to quality excellence, not only in styling but in utility, reliability, fuel efficiency, maintainability and safety, profit became the mark and measure of success and indifference about quality prevailed and a focus on lower cost became the driver of creative effort. Soon, even the designs, which were the counterbalance to lower under-the-hood value, began to grow lower in quality as increasing amounts of chrome over plastic replaced chrome over steel. As quality considerations lost prominence, it became easier for GM to build cars in foreign countries where politicians were eager to develop industry. In countries like Japan and China, labor was cheaper and it was also cheaper to operate manufacturing facilities because politicians and industrial leaders were unsophisticated in the long term dangers of industrial chemicals to workers and to the environment and they demanded few controls. These cost savings, and the relative immunity from the responsibility for the potential ill effects of manufacturing processes on foreign workers and their lands, more than offset the costs of building new overseas facilities and importing the finished products. As GM's profit grew, excessively, various conflicts with GM's American employees about wages, benefits, working conditions, and the export of their jobs escalated. These conflicts created an adversarial working environment where everyone sought to profit yet few focused on producing quality excellence.

When consumers realized that their American cars were, by design, intended to be obsolete in as short a time as possible, they increasingly purchased reliable, yet otherwise low quality, foreign cars. Over time, as our highways began to fill with foreign cars that were very old, by American standards, the reliability of foreign cars became undeniably evident and legendary. More importantly, the American consumer's concept of quality changed and made unreliable cars unacceptable. Lawmakers were eventually compelled by voter outrage to step in with "Lemon Laws" forcing car manufacturers to keep their customers' new cars running for at least three years – but this control was too little and too late. As one researcher noted "Detroit's quality nosedived just as import sales were growing. In 1960 imports had accounted for less than 5 percent of U.S. car sales, but by 1971 they accounted for about 15 percent..."[7] As the ability to charm customers through styling faded and reliability problems abounded, GM's unsustainable pocket of arrested culture was broken. As a result, leadership in sales and tens of thousands of jobs in the auto industry transferred to foreign nations – nations that pursued ascendant quality. Eventually, Toyota Motors of Japan became the world's leading producer of cars and GM went bankrupt because, even when GM changed tactics and built reliable cars, customers were too battle worn to trust them again.

Planned obsolescence became endemic as a business strategy, affecting many American consumer products, because the cost advantages and profit improvement over the more ethical, traditional, business practices made ethical businesses uncompetitive and impractical. This phenomenon was especially true for producers of complex products like cars where customers were unable to adequately assess and value better quality and safety as opposed to mere appearance. As manufacturers' qualms about production of shoddy products ebbed away, their pride in creating quality excellence was supplanted by an increasing acceptance of the idea that increasing profit, unrestrained by ethical or safety considerations – other than restraints imposed by law or by the costly repercussions in law suits by a handful of seriously injured customers – was the defining goal of a contemporary business and that profit, to the extent of greed, could be a source of pride.

This spirit of pride in greed was captured rather well as it was extolled as a virtue by Gordon Gecko, a fictional corporate raider played by Michael Douglas as the anti-hero in the movie Wall Street.

---

[7] Ingrassia, Paul. *Crash Course: The American Automobile Industry's Road to Bankruptcy and Bailout-and Beyond* (Random House, 2011) p. 49.

"Greed, for lack of a better word, is good. Greed is right. Greed works. Greed clarifies, cuts through, and captures, the essence of the evolutionary spirit. Greed, in all of its forms; greed for life, for money, for love, knowledge, has marked the upward surge of mankind..."[8]

Gordon Gecko
1987

While greed may have "marked the upward surge of mankind," greed has certainly not been the cause of it. On the contrary, greed, and its sociopathic[9] manifestations, have consistently prevented the upward surge of mankind. Greed differs from ethical risk taking which entrepreneurs manage in order to create new businesses or to create innovative improvements that have the potential to improve our quality of life in some way. Greed is embodied in excessive and unhealthy risks which introduce hazards, often at the expense of the many customers, for the good of the few or the one profiteer.

Although many businesses adopting planned obsolescence thrived for decades, consumers lost the benefit of improvements in products and services and in their quality of life. Our industrial base continues to pay the price in transfer of both jobs and ownership of the means of production to foreigners and foreign nations who naturally have their own best interests at heart. It appears that, even today, the mode of operation of business that Business Week magazine derided in 1956 is still business as usual.

"it is an age in which all the old admonitions appear to have been outdated: Make do. Neither a borrower nor a lender be. Penny-wise, pound-foolish. Waste not, want not. A penny saved is a penny earned. A fool and his money are soon parted. Just past the

---

[8] Weiser, Stanley and Oliver Stone. Screenplay: *Wall Street* (California: Oaxatal Productions, Inc., 1987)

[9] Readers interested in sociopathic manifestations of pride in greed may Google "The Business Plot" for details of a failed fascist coup to replace F. D. Roosevelt's government – which was seen as too favorable to employees at the expense of businesses. This plot was, allegedly, funded and planned by owners of GM and other conspiring big businesses. The plot failed when the retired Marine Corps general chosen to lead the coup was unwilling to participate and then testified against the conspirators. Also, see the biography of "Henry Ford" on the Wikipedia.com Web site for several examples of exceedingly perverse sociopathy.

midmark of the 20th Century, it looks as though all of our business forces are bent on getting everyone to do just the reverse: Borrow. Spend. Buy. Waste. Want."[10]

Business Week
1956

Planned Obsolescence and other sociopathic manifestations of pride in greed aren't in the exclusive domain of the once great companies; they are a persistent temptation towards an organizational culture that is adverse to ascendant quality, and to the progress of civilization, which can, in varying degrees, be created by almost any business at almost any time. The philosophy of pride in greed should be resisted by all decent businesses – suppliers and customers alike – as well as by the employees of businesses who bear the ability to improve and the common interest in a quality of life that improves over time.

It was GM's strategy of "a car for every purse and purpose" which defeated Ford's low cost strategy and it was Toyota's strategy of high reliability that defeated GM's planned obsolescence strategy. It should be noted that a strategy of high reliability was essential to foreign automakers, like Toyota, who had to contract for warranty repairs of their cars with their American competitors who had many service centers in every state. For this reason, the more unreliable foreign cars were, the more the foreign automakers' profit would be absorbed by expensive repairs, that were profitable to competitors, and the more incapable the foreign carmakers would become of survival in their foreign market. American automakers didn't have this lack of service capability in their foreign markets because, during WWII, GM and Ford worked very profitably with both the allied and axis powers and expanded their production and service facilities throughout Europe to meet the tremendous demand for cars, jeeps, trucks, tanks and even planes – all of which were needed in high quantity and not necessarily of high quality or high reliability. Alas, without the many production and service facilities in their foreign markets, which enabled the planned obsolescence strategy to flourish, it may have been possible for the American automakers to realize the need to revitalize their passion for quality before foreign competition

---

[10] *Business Week*, No. 1394 (McGraw-Hill, May 19, 1956) p. 202

began to capture their domestic market share. In this circumstance, the history of American automakers may have been very different.

The need for foreign automakers to exercise a strategy of high reliability for their cars, in order to survive in their foreign markets, can, to some extent, be confirmed by the fact that the reliability of Japanese and other foreign automakers' cars hasn't increased significantly, if at all, for decades, now that these foreign automakers own their own service centers in the U.S. The Toyota of today probably lasts no longer than the Toyota of the 1980's and the ability for customers to customize their new cars to their preferences has probably not improved either. Simple options that used to be common on American cars, like different styles of interior cloth, interior color, or type of stereo system are still unheard of on the most frequently sold foreign cars. Japanese and other foreign cars have not only ceased progress in reliability, and in other forms of quality, but they have also begun to show weaknesses in safety as a recent rash of serious safety defects, initiating very high profile recalls, shows. If cars made by Toyota actually were being continually improved we should expect to get a million miles out of each car by now with no recalls for safety defects. If the vulnerabilities of Toyota's current strategy sound like the vulnerabilities of Ford's strategy of the 1930's – that's because they are nearly the same – inflexible design and low creativity in providing an enjoyable product. Soon Toyota's aging high reliability strategy, which is now becoming a weak differentiator, as American cars continue to improve, will be defeated by a new strategy that better achieves quality excellence for the types of performance that customers desire. Whether this new strategy will emerge as a gradual, evolutionary, change or as a sudden, revolutionary, change remains to be seen. Whether Toyota recognizes and exploits the new strategy first or the baton of leadership in personal transportation transfers to another automaker also remains to be seen. From what we understand of the human spirit, driven by the quest for ascendant quality, this change is inevitable.

As this brief history shows, competition is a contest of strategies which are, in effect, a mixture of different forms and degrees of performance quality created by an organization and their products and services. Although technology has a lot to do with creating some forms of performance quality, and it may serve to significantly improve the degree of quality, competition is never a contest of technologies alone; but competition is always a contest among different strategies which encompass the broader spectrum of necessary performance. For these reasons, quality must be defined and understood in the context of its relationship with the customer's strategy, in pursuit of their mission, in order to ensure that the form and degree of performance offered matches the form and degree desired. In this way, the risk of poor performance, which is created when a business offering is inconsistent with the customer's concept of quality excellence, can be exposed and remediated before problems in the market take root. There is a one-best-way for an organization and their products and services to perform which is the one-best-way because the targeted customer group sees it that way as a supporting strategic component of the customer's pursuit of their mission. This fact mustn't go unnoticed in any effort to improve quality. The systematic pursuit of quality excellence is a competency that has the potential effect of creating an organizational culture that can be a breeding ground for the innovation that creates new and greater levels of success and profitability for most organizations. The capabilities to understand quality excellence, in relation to our needs or in relation to the needs of customers, and to formulate strategy that best achieves quality excellence, are central to advancing commerce, achieving healthy, competitive, markets, and a thriving culture that is capable of accomplishing great things.

In 1962, President John F. Kennedy, in his State of the Union speech, televised for the nation to see and hear and for the media to communicate worldwide, presented our nation the challenge to perform a mission which he called "the most hazardous and dangerous and greatest adventure on which man has ever embarked."[11]

> "I believe that this nation should commit itself to achieving the goal, before this decade is out, of landing a man on the moon and returning him safely to the earth."[12]

> John F. Kennedy
> 5/25/1961

---

[11] John F. Kennedy, Address at Rice University on the Nation's Space Effort, September 12, 1962; http://www.jfklibrary.org/Research/Research-Aids/JFK-Speeches/Rice-University_19620912.aspx

[12] John F. Kennedy, Speech to Congress on "urgent national needs," 5/25/1961; http://www.jfklibrary.org/JFK/JFK-Legacy/NASA-Moon-Landing.aspx

President Kennedy explained the incentive for this great adventure, to the students of Rice University.

> "We choose to go to the moon in this decade and do the other things, not because they are easy, but because they are hard, because that goal will serve to organize and measure the best of our energies and skills, because that challenge is one that we are willing to accept, one we are unwilling to postpone, and one which we intend to win..."[13]
>
> John F. Kennedy
> 9/12/1962

The conditions which created the Apollo 11 mission consisted of a coincidence in aims and a sense of purpose that have rarely occurred on such a grand scale in human history. President Kennedy began this endeavor with the incentives, firstly, to win the "space race" against the Soviet Union to the moon – a race which was competitively inspired by Soviet Cosmonaut Yuri Gagarin's, 1961, orbit of the Earth in the Vostok 1 spacecraft – and, secondly, to enjoy whatever benefits in the advancement of science and technology would come from it. Expectations with regard to these incentives were high and the motive to pursue them was expressed in the intentions and goals of the presidents' public strategic communication. These elements are the hallmarks of the conceptual system we call strategy – which can be found in all creative efforts. Moreover, the inspired development of new technologies and new strategies that use technologies for their best advantage are the hallmarks of life in a world on the quest for ascendant quality. The formidable challenges for which new technologies, and new strategies using them, were needed were given perspective by President Kennedy as he continued to expound on the mission to the moon in his speech to the students of Rice University.

> "...we shall send to the moon, 240,000 miles away from the control station in Houston, a giant rocket more than 300 feet tall, the length of this football field, made of new metal alloys, some of which have not yet been invented, capable of standing heat and

---

[13] John F. Kennedy, Address at Rice University on the Nation's Space Effort

stresses several times more than have ever been experienced, fitted together with a precision better than the finest watch, carrying all the equipment needed for propulsion, guidance, control, communications, food and survival, on an untried mission, to an unknown celestial body, and then return it safely to earth, re-entering the atmosphere at speeds of over 25,000 miles per hour, causing heat about half that of the temperature of the sun"[14]

John F. Kennedy
9/12/1962

It was our nation's shared vision that enabled NASA to orchestrate the vast ensemble of scientists and engineers and professionals, in many disciplines, to work together to harmoniously perform the many feats of developing and integrating technologies with strategies to resolve each challenge comprising the mission to the moon. The results of this effort bear reflection for what they tell us about the potential of our inspired capabilities. In a scene reminiscent of a science fiction movie, the world watched, as, moment after moment, a carefully scripted mission unfolded:

"ARMSTRONG: Houston, this is Neil. Radio check.

McCANDLESS: Neil, this is Houston. Loud and clear. Break. Break. Buzz, this is Houston. Radio check, and verify TV circuit breaker in.

ALDRIN: Roger, TV circuit breaker's in. And read you loud and clear.

ARMSTRONG: I'm at the foot of the ladder. The LM [Lunar Module] footpads are only depressed in the surface about 1 or 2 inches, although the surface appears to be very, very fine grained, as you get close to it. It's almost like a powder. (The) ground mass is very fine.

ARMSTRONG: I'm going to step off the LM now. (Long Pause)

ARMSTRONG: That's one small step for (a) man; one giant leap for mankind."[15]

---

[14] John F. Kennedy, Address at Rice University on the Nation's Space Effort
[15] NASA Broadcast, Neil Armstrong, Bruce McCandless II, and Edwin "Buzz" Aldrin, 7/21/1969

Neil Armstrong, Bruce McCandless II, and Edwin "Buzz" Aldrin
July 21, 1969

The Apollo 11 missions' successful moon landing and successful return to earth are great testimony to the capabilities of our civilization to expand the province and quality of life beyond their prior bounds when ascendant quality is a shared goal. Under the influence of an ascendant and shared goal, pride in greed and other human failings do not disappear – they are just less controlling influences than the shared desire to succeed in a challenging mission that has the potential to improve our quality of life. In this circumstance, the natural gifts of people arise to defeat the risk of poor performance wherever it appears. I'm introducing the term "Coascendancy[16]" to describe the condition of pursuit of the quest for ascendant quality as a shared goal as this condition was exemplified by the Apollo 11 mission. It's important for our language to have a term to describe the concept of Coascendancy in the hope that as we grow familiar with this term, and the desirability of the capability for social enrichment and cultural progress it represents, we may begin to envision, and communicate, ways to achieve Coascendancy in our shared endeavors.

As we evoke the silent echo of the memory of the first small step mankind has gingerly taken on our nearest celestial body, we cannot help but realize that mankind has not ceased to take small steps, moment after moment, with which the betterment of our quality of life has unfolded, in progress which amounts to a continuous series of giant leaps for mankind. It is this empirical evidence that supports the philosophical premise, stated in the preamble to this chapter, that "wherever life is found, life is found on the quest for ascendant quality." Not all life matches mankind's ability to integrate technology with strategy to attain incentives – but all life is incentivized to apply the technology it can master, with the strategies it can conceive of, to sustain survival, to improve its quality of life, and perhaps to secure a better future for its descendants. If, as this mounting evidence suggests, pursuit of the quest for ascendant quality is a fundamental need, which drives the behavior of individuals, and, consequently, the progress of our society, it is appropriate, if not essential, to base our understanding of how

---

[16] The term "Coascendancy" is defined, and further explored in chapter 5, The Ascendant Theory of Motivation.

quality is created and improved in our world on the basis of this philosophical premise.

The Process Enrichment Philosophy

The Process Enrichment philosophy, which our preceding discussion has introduced, is the basis of a disciplined approach to improving the human condition by using insights to the condition of us humans in order to do it. As our predecessors footprints, etched in the trail of our progress over the long course of history show, on whichever celestial body we may be found, the condition of us humans is that each of us is motivated by the inherent drives that compel us to use our minds and bodies to sense and discern quality excellence, in relation to the needs of the strategies that we pursue, for the incentives we desire, guided by the enlightening influence of risk, which continuously reminds us of the variance between our behaviors and those behaviors and better tendencies necessary to achieve quality excellence in everything we do. It's this basic model of the condition of the human, our world's form of life with the most prodigious and diverse creative ability, around which our evolving wisdom and philosophy are centered.

The Process Enrichment philosophy challenges us to risk openness in consideration of new perspectives on quality that bear the potential to transform and evolve our thinking and our personal connection with concepts that we live with on a daily basis beyond conventional wisdom. This challenge has all the rewards of exploration including the moment of truth when an epiphany occurs and the wisdom of the new philosophy becomes clear. This challenge and manner of progressively understanding philosophy was described well, nearly a century ago, by the philosopher Ludwig Wittgenstein.

"Philosophy is like trying to open a safe with a combination lock: each little adjustment of the dials seems to achieve nothing; only when everything is in place does the door open."[17]

Ludwig Wittgenstein
1930

---

[17] Wittgenstein, Ludwig. Maurice O'Connor Drury, "Notes on Conversations with Wittgenstein", in Rush Rhees (ed.), Ludwig Wittgenstein: Personal Recollections (Rowman & Littlefield Pub Inc., Totowa, NJ, 1981), p.126

The safe we are about to open has a combination lock with three very important and interrelated concepts as its tumblers – these are strategy, quality, and risk. Each of these tumblers must be dialed in and must work in simultaneous combination to open the door to the Process Enrichment philosophy.

As we work towards dialing in our understanding of the way the concepts of strategy, quality, and risk are interrelated we should find comfort in the realization that the ability to understand and balance strategy, quality, and risk is an inherent capability of all living beings. Proof of this is evident in the fact that if living beings can't achieve success in a strategy to find food, life would cease to exist; if living beings cannot manage to obtain the right form and quality of nutrients, life would cease to exist; and if living beings took risks or gambled with survival, life would eventually cease to exist. In each case, the quest for ascendant quality explains the behavior of living beings in seeking the strategy that will achieve the highest success in attaining the necessities of life at their highest quality and with the lowest risk of poor performance possible. The fact that balancing strategy, quality, and risk is a capability of all living beings, dating back to the origin of life, millions, if not billions, of years ago, is a testament to the essential value of these concepts in our ability to survive and to succeed in our efforts to create quality excellence and improve the human condition.

The significance of the interrelationship between the concepts of strategy, quality, and risk to our ability to achieve quality excellence, and to improve the performance of our strategies, is that as we understand the way these three concepts are interrelated we can understand how changes in one concept will create predictable, and therefore controllable, ramifications for the other two. Consider, for example, the act or strategy of purchasing milk from a supermarket to prepare for a breakfast of cereal. Milk is an item that is usually found with an expiration, or "sell-by," date stamped on the container. If the milk is purchased before the sell-by date all is well, however, milk doesn't necessarily spoil right away so, if there is only one container of milk left in the supermarket, and it's on its expiration date, a customer may purchase it – but the risk of poor performance of the customer's strategy – of the milk going sour when the customer needs it – travels with the risky milk wherever it goes. For this reason, even if the milk turns out to be fine, it is lower in quality because the customer's enjoyment of the milk wasn't ensured. In fact, the customer may decide on the way home from the supermarket to pick up another container of milk, at a different store, because the customer enjoys a good bowl of cereal and is willing to pay extra to shed the risk of poor performance from his or her strategy. Products that are of good quality shouldn't need to be purchased twice. As our own experiences tell us, for milk or for anything else, whenever risk increases, quality decreases; and whenever quality decreases from a level of quality excellence, strategy becomes less feasible or less productive and risk increases. Conversely, whenever quality is excellent, there is an absence of risk and strategy is gainfully performed. As we also know, whenever presented the opportunity to select from competing strategies, everyone can be depended on to select the strategy that they perceive will best serve their quest for ascendant quality.

As our discussion is increasingly making clear, there is a level of understanding of the concepts of strategy, quality, and risk which can only be reached by abandoning some of the ideas about these concepts, popularized during the last century's quality revolution, that we have become all too comfortable with. Some of the rationale for abandoning and growing beyond these ideas is summarized in the next few paragraphs. This brief retrospective is necessary to set the stage for us to begin to establish a new foundation, made ever stronger by the cohesion of a strong philosophical base, and our new awareness of the strong interrelationships joining strategy, quality, and risk together. Our new foundation will also build on a broader understanding of commerce as we redefine the process of commerce to achieve a more holistic context for illustrating the significance of strategy, quality, and risk to our commercial relationships and, ultimately, to the progress of civilization.

Strategy

Strategy is a topic for which countless books and articles have been written, over countless years, yet there is little consensus among the academic, military, and business communities about what the word "strategy" actually means. This fact was stigmatized, not long ago, by The Economist magazine.

"Nobody really knows what strategy is!"[18]

The Economist
1993

Contemporary society, our academic journals tell us, has derived the word strategy from the Greek word for generalship. This is, in a way, unfortunate because generalship is only one, very limited, application of strategy. The predominant uses of strategy don't involve defeating an enemy or gaining competitive advantage over another business. Strategy, as we discussed earlier, can be seen to operate in such commonplace planning processes as the planning of a breakfast of milk and cereal. As we can appreciate, since the dawn of life in our universe, life forms have almost continuously used strategy to attain incentives, such as nutrients and shelter, among other things necessary to survive and to thrive. Because of our own long history in the use of strategy, the cognitive processes and behaviors of living beings, such as ourselves, offer the deepest insight to the patterns and elements which can be used to properly model and define strategy.

The condition of us humans is that we have a keen natural ability to understand strategy when we experience it and we have a keen natural ability to spontaneously create strategy when we need it. Our natural ability to understand strategy, and the pattern of strategy, is perhaps best exemplified by the way we use our understanding of strategy in our justice system when we decide who in our society *is* and who in our society *is not* a criminal. As we know from our discussion in this chapter, the purpose of strategy is to attain incentives that support or advance our quest for ascendant quality. It is this purpose that is most telling when we are assessing another person's behavior. For example, if a man holds up a convenience store and walks out with a bag of potato chips and starts eating them we will put him in jail for a good while because we can readily understand that the man's incentive was to have a snack and the proof of the criminal strategy was in the behavior of eating the snack – yet, if the man runs out of the convenience store and throws the potato chips around like confetti we will be unable to recognize this behavior as a known pattern of strategy. In this case the man will be seen as "mentally ill" rather than as a criminal who is enacting a criminal strategy. It's the man's disregard for the incentive – the snack

---

[18] The Economist, March 20, 1993, p. 106.

of potato chips – that shows his behavior to be devoid of reason. Here we can see that the ability to understand strategy is a thread that runs deep in our wiring and can often serve to help us decide who among us is locked in a small jail cell and who is instead given medical care for roughly the same behavior.

Strategy, which is so vital to all that we do, and is so influential in how we view each other and the world, and is so instrumental in our creative processes, remains poorly defined – even though it is well understood intuitively. One consequence of this circumstance is that, while manufacturing and service organizations can often identify many customer needs, there is a deficiency in the ability to recognize their customers' overall strategy – often resulting in incomplete or inappropriate requirements and, consequently, in products and services with poor performance. The prevalence of problems on defense programs caused by the inability to adequately understand customer strategy is duly reflected in a cautionary note expressed in the Army's mandatory training for systems engineers.

> "When things go wrong on defense programs, the problems are often traceable to the failure to fully grasp the fundamental needs of the user."[19]
>
> Acquisition 201A
> Defense Acquisition University

Achieving a strong knowledge of customer strategy is not only a great way to fully grasp the fundamental needs of customers, and thereby avoid the things that often go wrong on defense programs, but it is perhaps the one best way to proactively enable the design and creation of better products and services faster and cheaper. As our discussion of strategy progresses, we will find that there is a new definition of the conceptual system we call strategy that is consistent with the way we use the concept of strategy in the courtroom, the boardroom, as well as in the war room. For this reason, only one definition of strategy is needed for these and all other applications. Also, because our new definition of strategy emulates a human modus operandi it enables us to comfortably bring to bear our

---

[19] Lightsey, Robert H. ACQ 201A *Intermediate Systems Acquisition, Part A*, 9/26/2011; A Program Manager's Guide to Systems Engineering, April 2005.

natural talents in recognizing and using strategy's elements and patterns in ways that benefit our creative effort.

Quality

Quality, like strategy, is also a concept that has, historically, been inadequately defined and has, consequently, contributed less than it should to our ability to improve our creative effort. The definition of quality still in use in the International Organization for Standardization's ISO 9000 series, and also in the Software Engineering Institute's Capability Maturity Model Integration for Development (CMMI-DEV)[20], albeit with slight variations, is the same definition of quality popularized by Philip Crosby who was a prominent leader in the quality revolution of the 1970's and 1980's.

> "Quality has to be defined as *conformance to requirements*. This definition places the organization in the position of operating to something other than opinion and experience. It means that the best brains and most useful knowledge will be invested in establishing requirements in the first place."[21]

> Philip Crosby
> 1984

The belief that the definition of quality is "conformance to requirements" is, in a very limited sense, useful in manufacturing and production line operations. In these industrial applications there is typically a strong desire on the part of management to eliminate creativity and prevent people from using their "opinion and experience" to do anything that in any way deviates from the efficient and repeated production of identical products. I learned of an interesting case of this when I worked as a quality engineer at Lockheed Missiles and Space Company, in the 1980's. At Lockheed, legend had it that once a product was designed, engineers were either let go or assigned to different projects. In this way, the potential for engineers to introduce new requirements or designs, regardless of whether they helped or harmed the end product, would

---

[20] Currently, the CMMI-DEV, ISO 9000, and Motorola University's Six Sigma Dictionary do not define strategy or risk and they do not describe the interrelationship between strategy, quality, and risk.

[21] Crosby, Philip B. *Quality Without Tears: The Art of Hassle-Free Management* (McGraw-Hill, 1984) p. 60.

be eliminated. Without changes in requirements and design, the product cost and production schedule were, presumably, closer to the approved plan. While the conformance to requirements definition of quality serves well in curtailing the creative thought process, and perhaps gives production employees enough information to produce something, the question remains: "Is conformance to requirements a true definition of quality?"

Ironically, the answer to this question can be found with Mr. Crosby's own words when he intended to define quality. By defining quality as conformance to requirements, this necessarily calls for the requirements to already exist in some form – but if the creative effort in developing the requirements is already complete then the only purpose for conformance to requirements is to control quality – not to define quality – quality must have already been defined by the "best brains" who developed the requirements. These "best brains" received no benefit at all from the conformance to requirements definition of quality. This observation shows how the logic of the conformance to requirements definition of quality breaks down. Testing the logic of the "quality means conformance to requirements" definition of quality we can quickly realize that it begs the question "How do you define quality for the requirements?" There simply is no way to determine the quality of requirements when the definition of quality is "conformance to requirements." This is the essence of the conformance to requirements fallacy. As this retrospective analysis shows, the conformance to requirements definition of quality was never intended to help the creative effort itself; rather, it is a convenient expedient, intended to serve as a control mechanism to assure smooth operation of an effort in production. This is a worthy cause, but it is unrelated to the essence or elements of quality.

A true definition of quality would help us to become the experts with the "best brains and most useful knowledge," that Mr. Crosby spoke of, who serve in "establishing requirements in the first place." The "quality is conformance to requirements" definition doesn't enable us to do this – and it wasn't intended to. In fact, conformance to requirements is the antithesis of the definition of quality needed to support a creative effort – and, as we should realize, virtually all effort begins as creative effort. In today's world, where software development and business services of all kinds occupy most of us, and manufacturing industries continue to move overseas, the "conformance to requirements" definition of quality has lost the great part of its relevance and usefulness.

The legacy of the "conformance to requirements" definition of quality is that in contemporary society, as this definition of quality continues to be adopted for use in non-manufacturing endeavors, it increasingly exerts a detrimental impact on our culture as it serves to inhibit our natural creative drives and acts to displace responsibility for creating quality to someone else within the organization who is responsible for defining requirements. This displacement enables people to sometimes continue, over and over again, to create products that not only perform poorly but are sometimes actually dangerous to customers – just because the products conform to requirements. After all, it's enormously difficult to fight a system that tells us, and pays us, to do as the requirements tell us to do. In this regard, conformance to requirements serves as a protective shell that excuses us for any harm caused by following instructions – and we acquiesce in diminished quality and excessive risk rather than confront the feared storm of possible repercussions coincident with standing up for quality excellence.

The realization that there is a "conformance to requirements fallacy" is proper foundation for understanding the nature of the potential for great harm inherent in the "conformance to requirements" definition of quality. In the case where requirements are unsuitable or inadequate to the purpose of a service effort or to the function of a product, an unacceptable risk of poor performance is introduced – sometimes with dangerous consequences. This potential for great harm is exemplified by Mr. Crosby's use of his definition of quality to describe the quality of one of the Ford Motor Company's most notorious cars.

> "If a Pinto conforms to all the requirements of a Pinto, then it is a quality car."[22]
>
> Philip Crosby,
> 1980

The Ford Pinto was a car that conformed to all the requirements of a Pinto – so it met the definition of quality espoused by Mr. Crosby. Unfortunately

---

[22] Crosby, Philip B. *"Quality is Free: The Art of Making Quality Certain,"* 1980.

for buyers of Ford's Pinto, the requirements for the Pinto – which were inadequate with regard to safety – assured that the Pinto would never be a quality car no matter how precisely or skillfully the most talented manufacturing personnel, with the latest technology, worked to produce it. An article in Mother Jones magazine provides details.

> "Ford engineers discovered in pre-production crash tests that rear-end collisions would rupture the Pinto's fuel system extremely easily...
>
> Ford waited eight years [to address Pinto's safety hazards] because its internal "cost-benefit analysis," which places a dollar value on human life, said it wasn't profitable to make the changes sooner...
>
> When it was discovered the gas tank was unsafe, did anyone go to Iacocca [Lee Iacocca, president of Ford] and tell him?
>
> "Hell no," replied an engineer...
>
> "That person would have been fired. Safety wasn't a popular subject around Ford in those days. With Lee it was taboo. Whenever a problem was raised that meant a delay on the Pinto, Lee would chomp on his cigar, look out the window and say 'Read the product objectives and get back to work.'"
>
> ...As Lee Iacocca was fond of saying, "Safety doesn't sell."[23]
>
> Mother Jones
> 1977

The tragic and unnecessary loss of life inflicted on Pinto customers offers a stark example of the consequences of misuse of the conformance to requirements definition of quality. In this instance, Mr. Iacocca used his demand for conformance to requirements in the "product objectives" as an instrument of authority to displace responsibility from employees and enable them to suppress their objections to participation in building for customers the car that exemplifies the word "deathtrap."

---

[23] Dowie, Mark. *Pinto Madness* (Mother Jones, September/October 1977 Issue), http://www.motherjones.com/politics/1977/09/pinto-madness

The potential for great harm in this obedience to authority dynamic – although under different circumstances – was experimentally validated in Yale University's landmark study by Dr. Stanley Milgram.

> "Ordinary people, simply doing their jobs, and without any particular hostility on their part, can become agents in a terrible destructive process. Moreover, even when the destructive effects of their work become patently clear, and they are asked to carry out actions incompatible with fundamental standards of morality, relatively few people have the resources needed to resist authority. A variety of inhibitions against disobeying authority come into play and successfully keep the person in his place."[24]
>
> Dr. Stanley Milgram
> 1974

This ability of the "conformance to requirements" definition of quality to enable normal, well educated, people to disengage from their professional and moral responsibilities isn't an isolated case – rather it's, in varying degrees, the expected case. It's typically very difficult for employees to challenge decisions already made and codified into requirements when they recognize that something has gone wrong. It's also likely that the normal channels for addressing defects may resist the attempt to label something that conforms to requirements a defect. This is significant to consider in understanding the nature of the impact of a definition of quality on organizational culture.

My most relevant personal experience with this dynamic was an occasion when I was newly hired as the quality manager for a large sheet metal factory. I visited the shipping dock and noticed that there were stacks and stacks of finished goods piled up. I was told various sordid stories about the customer and how picky they were – with the upshot being that their receiving inspector wouldn't buy my employers' products even though the products conformed to the customers requirements. I confirmed that the "door panels" – these were sheet metal doors for a computer system – were made in the usual manner and met requirements. I then joined the customer's receiving inspector as he reviewed

---

[24] Milgram, Stanley. *Obedience to Authority: An Experimental View*, 1974

a fresh lot of Door Panels. As it turned out, even though the Door Panels met requirements, they weren't adequate for use in a Clean Room as the customer intended. The industry standard glue used in production had dried out and became dusty from the heat of the painting process that baked the paint. There were also spots where the electroless nickel plating was thin. Although profit margin was good on these Door Panels, and several hundred thousand dollars worth of material was tied up on the shipping dock, it didn't occur to the production personnel to simply address the customers concerns and keep this large customer happy. All it took was cleaning with Q-tips and a little metallic spray paint to make all well.

Quality professionals see this type of dynamic in organizations frequently. We encounter production personnel who become perplexed by situations where a product that would be acceptable to most customers is unacceptable to a few customers. This undoubtedly occurs because once conformance to requirements has been achieved the ability and even the desire to adapt, innovate, and promptly solve simple problems often breaks down. At that point, production needs an intervention by someone, perhaps armed with statistics, to help them snap out of it.

In organizations which adopt the fallacious belief that "quality is conformance to requirements," it can become rather easy, in the interest of quality improvement, to expand the scope of that belief to also consider organizational "processes" to be requirements. It is through this progression of thought that we find the probable cause of the "quality is conformance to requirements" fallacy extending its scope of influence into a belief that the Software Engineering Institute (SEI) calls "the process management premise."

"The SEI has taken the process management premise, "the quality of a system or product is highly influenced by the quality of the process used to develop and maintain it," and defined CMMs [Capability Maturity Models] that embody this premise. The belief in this premise is seen worldwide in quality movements, as evidenced by the International Organization for Standardization / International Electrotechnical Commission (ISO/IEC) body of standards. CMMs focus on improving processes in an organization."[25]

CMMI® for Development, 2010

The process management premise creates the fallacious inference that *conformance to improved organizational processes will lead to improved product and service quality for customers.* This fallacious inference is also conveyed in the subtitle of the Capability Maturity Model Integration for Development (CMMI-DEV) which reads "Improving processes for developing better products and services."[26]

As you may confirm, the only process that can improve product or service quality is a quality improvement process. It is only to the extent that a process improvement also includes a quality improvement that an organization will find improved quality in their products and services. As shown in Table 1-1, quality improvements are primarily intended to improve the market value and performance of a product and service while process improvements are primarily intended to improve the quality of an organization. One advantage of recognizing what we can refer to as "the process management premise fallacy" is that this recognition clarifies that, although they are both inspired pursuits, quality improvement and process improvement are not interchangeable pursuits. If you have ever been to a fast food restaurant which has consistent, efficient, and well managed processes but serves unhealthy, undercooked, food, that you promise yourself you will never eat again, you can realize that improving the performance of the processes of an organization and improving the performance of the products and services an organization produces are two very different activities with very different purposes requiring very different capabilities and very different applications of reasoning and talent from which very different customers derive very different benefits.

---

[25] CMMI® for Development, Version 1.3: Improving processes for developing better products and services, 2010
[26] CMMI® for Development, Version 1.3

QUALITY IMPROVEMENTS	PROCESS IMPROVEMENTS
An affordable new feature that customers want is added.	A process now creates fewer defects.
A more effective technology is provided; i.e. better miles per gallon, faster acting pain medicine, etc.	Transferring the customer support call center to an English speaking country, with lower labor costs, improved profit.
A product is built using materials that last longer.	Longer lasting lubricant is used which increases productive time.
A software application can now correct spelling errors automatically.	Scrap material is reduced.
A chair uses material that's more comfortable to sit on.	People can find out how to get things done in the organization more easily.
Service is offered in multiple languages to assist customers worldwide.	A process that was too complicated can now be done easily and by more people.

Table 1-1; Examples contrasting quality improvements and process improvements.

Quality improvements, which customers experience, arise from an almost entirely different, customer focused perspective than from the perspective necessary for organizational process improvement, yet, the phenomenon of managers paying disproportionate attention to process improvement and process audits rather than to improvement of the ability of their products and services to help customers enact their valued strategies is plain to see in far too many organizations. Astute leaders are beginning to take notice of this phenomenon. Mark Schaeffer, the Department of Defense sponsor for CMMI and Director of Systems Engineering for the Office of the Secretary of Defense, delivering a keynote address at a CMMI Technology Conference and User Group, said, as reported by Mike Phillips of the Carnegie Mellon University:

"When achieving a [CMMI Maturity] level replaces the focus on continuous improvement, we've lost sight of the goal." He held that appraisal results are too often an end in themselves instead of meeting the government's expectations for actual performance on complex, software-intensive system development."[27]

Mark Schaeffer

A reasonable explanation for this compulsion to conform to requirements and conform to processes and to comply with process models written by independent authorities, often regardless of cost, in order to achieve prestigious Maturity Levels, at the expense of effort better spent perfecting the quality and competitiveness of an organizations products and services, is provided by Dr. Thomas Blass, an authority on Dr. Milgram's obedience research.

"Once you accept another person's authority, you become a different person... You are concerned with how well you follow out your orders, rather than whether it is right or wrong."[28]

Dr. Thomas Blass

The pursuit of assessment at higher and higher Maturity levels would seem to be an ideal indicator of how well an organization has accepted the CMMI's authority and has followed out the CMMI's processes, rather than whether those processes are producing products or services that customers want. It is this form of obedience, and disengagement from the responsibility to create valuable, competitive, products and services, that is the great danger introduced by fallacies and misconceptions that obscure the meaning of quality.

There are at least two types of process improvement methods which are commonly believed to be quality improvement methods even though they possess no significant ability to improve quality. Firstly, there is the belief that a process

---

[27] Phillips, Mike. *CMMI–A Progress Report* (Software Engineering Institute, Carnegie Mellon University, 1/1/2005). Retrieved on 5/31/2014, http://www.sei.cmu.edu/library/abstracts/news-at-sei/cmmiinfocus20051.cfm

[28] Wolfson, Andrew. *A hoax most cruel: Caller coaxed McDonald's managers into strip-searching a worker*, (Louisville Courier-Journal 10/9/2005) Retrieved on 5/31/2014, http://archive.courier-journal.com/article/20051009/NEWS01/510090392/A-hoax-most-cruel-Caller-coaxed-McDonald-s-managers-into-strip-searching-worker

that reduces, eliminates, or fixes defects improves quality. Obviously, a product that *is not* defective is of better quality than one that *is* defective; however, since most organizations are not in the business of selling defective products, all the benefit in the time, effort, and cost savings from eliminating sorting and rework accrues mainly to the productive organization – not to the customer. Also, even if a business reduces defects to the extent of 3.4 defects per million opportunities, as is the custom in Six Sigma, this only maintains a product or service in the condition the designer planned to attain in the first place – no new functions or abilities were added, no increase in market value occurred, rather, for customers, the product or service is the same as it ever was. Secondly, we can recognize that, in the same way repairing defects doesn't improve quality, Statistical Process Control (SPC) techniques, which raise warning signs when a production process exhibits a worsening trend, do not improve quality either. The basic tools of quality including SPC charts, Ishikawa diagrams, Pareto analysis, scatter charts, checklists, etc. can be used for quality control and quality assurance and perhaps for process improvement – but these tools, cannot, but indirectly, if at all, produce quality improvement. In the absence of a proper definition of quality, industry has been stretching and reaching for what they cannot grasp without letting go of the fallacious and obsolescent beliefs of the last century.

There is one last misconception about quality that will be worthwhile to expose. This misconception may have taken hold as far back as 1901, when Ransom Olds invention of the assembly line concept enabled his Olds Motor Vehicle Company to become the first company in America to mass-produce automobiles. With mass-production, Olds was able to increase productivity 588% – from "425 cars in 1901 to 2,500 in 1902."[29] Nowadays, as we try to learn from the highly successful productive philosophy of Toyota, the term "Lean manufacturing" came into vogue and the association with quality endured, since Toyota was rapidly producing cars of relatively high quality. On Toyota's Web site they describe the Toyota Production System (TPS) as:

> "A production system that is steeped in the philosophy of the complete elimination of all waste and that imbues all aspects of

---

[29] Redgap, Curtis (2007). "Pioneers of the auto industry". http://www.allpar.com/corporate/bios/pioneers.html. Retrieved on 2008-01-08.

production with this philosophy *in pursuit of the most efficient production method...*

Toyota Motor Corporation's vehicle production system is *a way of "making things"* that is sometimes referred to as a "lean manufacturing system" or a "Just-in-Time (JIT) system," and has come to be well known and studied worldwide...

Based on the basic philosophies of jidoka and Just-in-Time, the TPS can efficiently and quickly produce vehicles of *sound quality*, one at a time, that fully *satisfy customer requirements*."[30] [Emphasis added]

Toyota Motor Corporation
2010

Although, from the viewpoint of Toyota's philosophy, they are building cars one at a time, their actual production rate is a little faster than 13 cars per minute[31] – for every minute of every day of the year – or more than 19 thousand cars per day. An important point to note is that Toyota refers to its lean manufacturing system as "a way of making things" in other words – a strategy, with the stated goal of quickly producing vehicles of "sound quality." The benefit of "Lean" is in speed of throughput which, luckily enough, has a byproduct of encouraging the improvement of processes so that defects and rework, which can slow production, are minimized. In Toyota's own words, Toyota seeks "sound quality" for their cars – not quality excellence or perfection as some might expect. For this reason, we don't look to Lean as a means of defining or improving quality.

Any system intended to improve quality should have a valid and useful definition of quality to work with. The "conformance to requirements" definition of quality doesn't conform to this fundamental requirement. With the exception of some forms of repetitive production efforts, contemporary industries do not seek to sustain conformance to requirements – rather it is a sustained creative and innovative effort that contemporary industries seek to

---

[30] Toyota Production System, Retrieved in 2011, http://www.toyota-global.com/company/vision_philosophy/toyota_production_system/

[31] Calculated based on OICA,s 2008 data, http://oica.net/wp-content/uploads/toyota1.pdf (Retrieved 2010)

sustain – with thinking people integrating thoughtful creative effort in their work – based on their knowledge and talents. In software development, for example, since so much is left to the software engineers to decide, the term "derived requirements" was coined to describe requirements the software engineers develop on their own – often without customer concurrence. In some software applications the derived requirements may not be complete until the final build. In this case, once again, the inability of the conformance to requirements definition of quality to help people engaged in a creative effort to know what "good" means for their work is a serious shortcoming. Deep change in the way we define and understand quality is necessary if we are to pursue the path of ascendant quality with confidence and faith in success. As we will soon see, a correct definition of quality has the power to engage employees in the quest for ascendant quality and can transform leadership into a beneficial instrument connecting creative effort to the ends of organizational strategy.

Risk

Risk, like strategy and quality, and perhaps more so, is also a regularly misrepresented concept in the academic journals – which is made more disturbing because we all know risk all too well when it affects us. The challenge in defining risk is amply demonstrated by the NASA Engineering and Safety Center (NESC) in a report that they courageously released publicly.

"Following two days of discussions on risk definition, it was agreed that, Fell[32] encompassed the mood toward risk that the NESC currently has adopted:

"Unfortunately, there are no generally accepted definitions of the terms used in risk assessment …… shortly after its formation in 1981, the United States Society for Risk Analysis established a committee to define risk. After 3 or 4 years of work the committee published a list of 14 candidate definitions and reported that it could not reach agreement. They recommended that a single

---

[32] Fell, R. "Landslide Risk Assessment and Allowable Risk," Canadian Geotechnical Journal, 31(2): 261-272. 1994.

definition of risk not be established but that every one be free to define it as appropriate to his or her own work."[33]

NASA Engineering and Safety Center

Reading further, the NESC's report shows that they settled on the definition of risk in the Risk Management Guide for Department of Defense (DoD) Acquisition. The DoD is an excellent choice since their guidance supports this country's large defense industry. The definition of risk provided by the DoD was revised since the NESC's report was published to the current version shown below.

"Risk is a measure of future uncertainties in achieving program performance goals and objectives within defined cost, schedule and performance constraints.

... Risks have three components:

• A future root cause (yet to happen), which, if eliminated or corrected, would prevent a potential consequence from occurring,

• A probability (or likelihood) assessed at the present time of that future root cause occurring, and

• The consequence (or effect) of that future occurrence.

A future root cause is the most basic reason for the presence of a risk. Accordingly, risks should be tied to future root causes and their effects."[34]

Risk Management Guide for DoD Acquisition
2006

This version of the DoD's definition of risk added the idea that a component of risk is "a future root cause." This idea doesn't appear to me to be justified; on the contrary, all the evidence indicates that the root cause of any risk always happens coincidentally with the creation of the risk – not sometime in the future – that's why we say that the cause is at the root rather than somewhere higher up. Consider, for example, a gambling metaphor. The moment you drop a quarter into a slot machine you have created the root cause of the risk of losing

---

[33] Development of Risk Assessment Matrix for NASA Engineering and Safety Center, Kelly D. Moses. P. E., Futron Corp., Roy W. Malone, Jr. NASA Marshall Space Flight Center (Downloaded, 12/18/2011, from: http://ntrs.nasa.gov/archive/nasa/casi.ntrs.nasa.gov/20050123548_2005093494.pdf)

[34] Risk Management Guide for Department of Defense Acquisition, Sixth Edition, version 1, Aug. 2006

the quarter because you're exercising a strategy of gambling. The risk remains until you know if you've hit the jackpot. It wouldn't make sense to say that the wheels of the slot machine coming up without a match was the root cause of the risk – the risk is over at that point. Another disadvantage of expecting risks to have "a future root cause" is that instead of looking at the harm of a strategy itself, risk managers are distracted and are looking at only the rare events that would disrupt a strategy – regardless of the harm the strategy itself is actively causing – for example, risk managers often look at risks of late deliveries or risks of defects rather than risks of poor performance in the strategy itself such as impracticality, undesirability to customers, excessive waste, complexity, etc. – these are significant risks that a risk manager can actually do something about. Unfortunately, our institutionalized definitions of risk do not create organizational cultures where these forms of risk can be understood or communicated in an actionable way. Using a misleading definition of risk, which gives a pass to poor performance, may, in fact, be worse than not formally tracking risk at all.

The remaining two components of risk – the probability and the consequence of something going wrong – also fail to capture the meaning of risk. As we discussed in our example of purchasing milk on its expiration date, our hypothetical customer may have decided on the way home from the supermarket to go out of his or her way to buy another container of milk to shed the risk of the milk going sour when it was needed. This risk, regardless of whether or not the risk was realized, by the milk going sour, was a costly distraction because the risk was an influence on the customer's thoughts and behavior. This influence, which is risk's predominant characteristic, cannot be adequately captured as a probability or consequence and, therefore, cannot be understood using contemporary definitions of risk. As we develop and redefine the concept of risk in the next few chapters, the presence of the essential focus on risks influence, along with risk's effect on strategy and quality, will enable us to benefit from our deep experience in living with risk in applying this knowledge to the improvement of our creative efforts.

While we may not be as concerned about the risk of milk going sour as we are about the risks of multi-billion dollar defense projects, surgical procedures, or business ventures going sour – we should recognize that the benefit of a deep understanding of strategy, quality, and risk is that, since these concepts are pervasively present in the thought processes governing decisions relevant to everything we do, as we improve our ability to benefit from our knowledge of these concepts we can expect to realize benefits in all our decision making challenges. Also, as we define the concepts of strategy, quality, and risk in a way that captures their interrelationship, and also their meanings as concepts independent of their industrial use, we will have the ability to enjoy the greater freedom of quality improvement concepts with universally broad application to all things great and small.

Commerce

As our previous discussion highlighted, commerce is a process that is central to improving our quality of life. The alternative processes available for attaining the goods and services people need being warfare and anarchy, the value of sustained quality excellence in commerce becomes clear. While the processes of commerce change over time, under various economic and political drivers, and the various rules and regulations that evolve, in its essential form, commerce is simply a trade for mutual benefit. Given what we know of the human need to be on the quest for ascendant quality, we can expand this basic definition of commerce to create a definition that is, perhaps, more holistic and more useful to us in our efforts to achieve quality excellence in our commercial activity. The Process Enrichment philosophy provides an ascendant definition of commerce which serves this purpose.

> ### COMMERCE
>
> *Commerce is the process of the ethical and equitable trade of assets which each trading party expects will serve a strategy that they envision to be supportive of their quest for ascendant quality.*

In this new definition of commerce we can see that the principal motivation causing the buying and selling of goods and services is the coincidence of each party being attracted and incentivized to engage in trade by seeing the trade, at the time, as an acceptable way, or, ideally, the one-best-way of achieving progress in a strategy that they perceive to be able to support or improve their quality of life. In this light, commerce is dependent on and driven by Coascendancy between trading partners in which each party attains valued benefit because of their ability to provide valued benefit to the other party. This definition of commerce is consistent with the belief that mutual benefit and societal development are the characteristic achievements of ethical and equitable commerce. The qualifier that there needs to be ethical and equitable trade in order for the process to be called commerce differentiates the process of commerce from the processes of fraud, theft, or coercion which tend to result in unfair trades which are contrary to the dictates of civilized behavior and which are often reversible in the courts. Use of techniques such as planned obsolescence, or asymmetric warfare, where a company dominating a market exploits their power in ways that inhibit the improvement of quality in the products or services of their industry is still legal in commerce – despite the fact that this serves to create or perpetuate an arrested culture – and despite the fact that these companies don't openly express their relationship with customers as a belligerent one – regardless of how sociopathic their behavior becomes. As mentioned earlier, the best assurance that the risk of poor performance in commerce is at a minimum comes from the talent of the trading parties themselves in their ability to understand the elements of quality. This definition of commerce has been our first small step on our journey towards an improved talent in creating quality excellence as it serves to help us understand the significance of commerce in shaping our Coascendant relationships.

As our new definition of commerce suggests, customers do better in commerce as they become better skilled in identifying and obtaining the goods and services that provide the one-best-way to advance their strategy. The extent of a customer's success in selecting the best market offering is predominantly dependent on their abilities to:

- Define and Establish Strategy
  Customers should be able to envision and understand their own strategy in good detail.
- Define Quality Excellence
  Customers should be able to use their established strategy to define the form and degree of quality that balances sufficient benefit with tolerable risk.
- Evaluate and Select the Best of the Competing Strategies

Customers should be able to compare the competing strategies offered by the suppliers that serve the market and select the alternative that best achieves quality excellence.

Each of these abilities is, in a sense, a Customer's Screen through which their decision making information is filtered and judgments are made with regard to the various market offerings presented to them by their suppliers or potential suppliers. As these abilities are improved, it is likely that better acquisition decisions will result.

On the other side of a trade, a suppliers' challenge is to formulate a product or service offering, and a manner of organizational performance, for their customer market that offers a way to enact the customer's strategy that customers will find attractive – and ideally, the most attractive from among competing offerings. The basis of the customer's attraction is the absence of deviation of the supplier's offering from the customers view of quality excellence – or, in other words, customers will select the solution offering the minimum risk of poor performance. The formulation of an optimal offering to customers is predominantly dependent on the supplier's abilities to:

- Discern the Customer's Current Strategy
  Suppliers should be able to discern their targeted customer group's current or intended strategy for the product or service offering.
- Establish the Customer's Concept of Quality Excellence
  Suppliers should be able to define the strategy representing the form and degree of quality that balances sufficient benefit with tolerable risk that the targeted customer group equates to quality excellence.
- Formulate, Create, and Offer a Mutually Beneficial Strategy
  Suppliers should be able to formulate, create and offer a strategy that the targeted customer group can identify, from among competing strategies, as the one-best-way to affordably accomplish their mission. This strategy must also provide the supplier an appropriate profit commensurate with the risk, resources, and effort applied.

Deficiencies in these abilities can serve as Customer Assessment Barriers which can act as blind spots preventing the supplier's best efforts from attaining success due to an absence of the knowledge necessary to perform competitively and defeat the Customer's Screens. As suppliers improve these abilities, it is likely that better commercial offerings will result.

The basic elements of the customer to supplier relationship are shown in Figure 1-1.

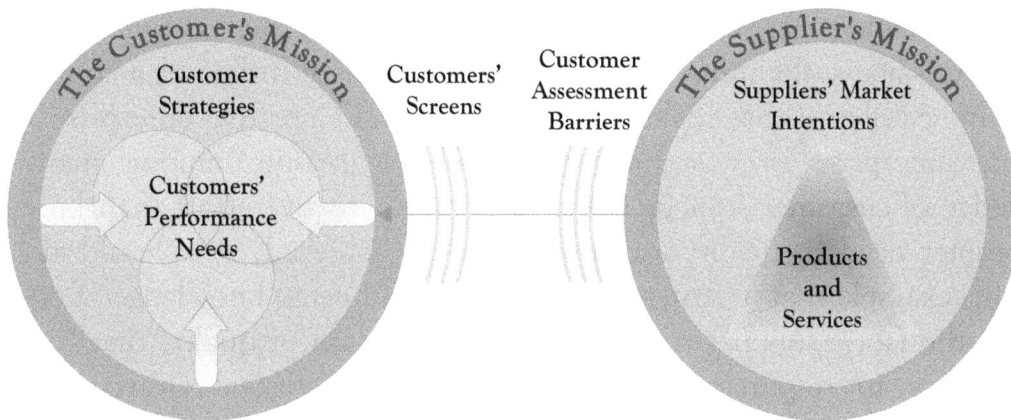

Figure 1–1; Basic Elements of the Customer to Supplier Relationship

In figure 1-1, we can imagine in the circle to the left that a customer is operating Customer Strategies through which the customer accomplishes their mission. Each Customer Strategy is comprised of the Customers' Performance Needs. In this example, some of these needs are common to various strategies. In order for the customer to select a supplier to meet some of the needs of a strategy, or to meet the needs of one or more entire strategies, the customer must be able to identify a supplier whose product or service can get through the Customer's Screens. If a supplier's market intentions are to serve the customer on the left, the supplier must overcome their own Customer Assessment Barriers and create products and services that most easily pass through the Customer's Screens.

The customer to supplier relationship, pictured in figure 1-1, is often a repeating sequence as customers become suppliers to their next higher customer. As an example, a company that manufactures fabric may sell rolls of fabric as a supplier to a clothing manufacturer, the clothing manufacturer may sell the clothing as a supplier to a designer, the designer may sell the clothing as a supplier to a retail chain and the retail chain may sell the clothing to end customers. If the fabric supplier, at the beginning of this repeating sequence of customer to supplier relationships knows that the end customer is wearing the clothing as pajamas the supplier won't try to sell their next higher customer flammable fabric – even if the customer forgets to specify "non-flammable fabric" in the requirements of the fabric order. This is why it is best to use talented suppliers even if it sometimes costs a little more. We should also note that this supplier's talent is derived from a deep knowledge of the customer strategy – several levels above the immediate customer. It is talent driven by a commitment to quality excellence that makes this manufacturer's creative enterprise a more valuable enterprise to customers. This is one reason, among many, that knowledge of customer strategy is more useful to creative effort than knowledge of customer needs if quality excellence is to be achieved. We should also note from this example that if, instead of possessing and using knowledge of customer strategy, the fabric supplier simply used the definition of quality that "quality is conformance to requirements," the lack of technological awareness of the customer, as indicated by their not specifying "non-flammable fabric" in their requirements, would have resulted in a harmful product, in this case, flammable pajamas, that probably could not be sold. In this example we can see that the fabric of quality excellence in commerce is woven from integrity and talent, driven by awareness of customer strategy, and a commitment to quality that serves to remove the risk of poor performance wherever it can be found – in a word, Coascendancy.

Understanding a customer's strategy is perhaps the best way to understand their acquisition behavior and the form of productive behavior necessary to win the customer's business. While a customer's stated requirements may be a good way to understand what a customer needs, understanding their strategy can fill in any knowledge gaps or even expose inappropriate or unnecessary requirements. Knowledge of customer strategy is especially important for products or services where the supplier is more knowledgeable of the relevant technology than the customer. It's with the customer's strategy that suppliers can gain the clearest picture of what their customer would like to do, how they currently do it, and how innovation can be introduced and quality improvement applied to advance the customer's mission and, for helping customers succeed in their strategy, suppliers can best attain the market incentives that they desire.

By understanding commerce as more than just the exchange of goods and services, we can now recognize the ageless benefit of understanding customer strategy and customer screens as well as customer assessment barriers. These classic features have undoubtedly accompanied commerce for thousands of years and will probably be relevant for thousands of years more.

It is in the pursuit of the truth of how the concepts of strategy, quality, and risk are defined and interrelate, and in the pursuit of the ability to create quality excellence in all that we do, and in the interest of Coascendancy in all that we do together, that we find the heart of the Process Enrichment philosophy and the origin of the Process Enrichment® methodology. Our quest is consistent with the ideals expressed by the famous French philosopher Voltaire.

"The pursuit of what is true and the practice of what is good are the two most important objects of philosophy."

Voltaire
(1694-1778)

The Quest for Ascendant Quality is a battle that will rage on as long as life seeks satisfaction though creative effort. For this reason it is essential to understand the elements of the great processes that create quality excellence in our world through philosophy that will stand the test of time with the support of a growing body of methodology that makes our philosophy practical.

The next waypoint on our journey to a deeper understanding of the concepts of strategy, quality, and risk runs through the context of Strategic Communication – to a place where we will consolidate our new understanding of these concepts and build new definitions that establish the Process Enrichment philosophy and that are consistent with the new knowledge we have already discovered on the quest for ascendant quality.

Chapter 2:

Elements of Strategic Communication & Quality

Strategic Communication is the active ingredient in an organization that harmonizes the inspiration and creative direction of the organization's guiding motives with the people who apply their creative effort for and the people who receive creative effort from the organization.

Chapter 2:

Elements of
Strategic Communication & Quality

PERSPECTIVES

- ▶ The definition of strategic communication
- ▶ How a leader's strategic communication acts to advance organizational strategy, improve quality, productivity, and value to customers
  - ▪ The Coral Sea Principle
- ▶ The patterns and elements of successful strategic communication
- ▶ Quality's six Themes of Performance
- ▶ The definition of quality
- ▶ The definition of strategy
- ▶ The Ascendant Strategy Model
- ▶ Formalizing and communicating organizational strategy using the Balanced Enterprise Strategic Themes (BEST) framework

Strategic communication has been a necessary and valuable part of leadership and the organization of people in various productive or even destructive enterprises since before the earliest days of recorded history. Often enough, if we search political history, military history or corporate history to find a great leader we will have also found someone with mastery of strategic communication. Leaders with mastery of strategic communication have a talent that serves to rally people behind the principles and purposes which serve to drive the creative direction of the societies in which we live and the businesses in which we work. Strategic communication is an essential part of business life because no service can be delivered and no product can be produced without a strategy to guide the effort. For this reason, the quality of strategic communication is a significant determinant of the quality of the products and services an organization is capable of producing. Strategic Communication is, also, a vital means of creating market demand for the fruits of an organizations' creative effort.

One of the best developed definitions of strategic communication is provided by the Joint Chiefs of Staff of the U.S. armed services.

> "Strategic communication (SC) refers to focused [United States Government] USG efforts to understand and engage key audiences to create, strengthen, or preserve conditions favorable for the advancement of USG interests, policies, and objectives through the use of coordinated programs, plans, themes, messages, and products synchronized with and leveraging the actions of all instruments of national power. SC combines actions, words, and images to influence key audiences."[35]

> Joint Chiefs of Staff
> JP 5-0

This Joint Chiefs of Staff definition of strategic communication easily translates to a generic definition, applicable to any organization, by substituting the word "organizational" for the words "United States Government" or

---

[35] Joint Publication 5-0, Joint Operation Planning, 11 August 2011, http://www.dtic.mil/doctrine/new_pubs/jp5_0.pdf

"national." Usage of the term "strategic communication" is nearly identical in its political, military, and organizational applications. For this reason, experiences in politics and warfare, which are usually analyzed thoroughly by historians, can provide reliable lessons demonstrating the elements of strategic communication that make strategic communication successful.

In organizations of all sizes, leaders seek to craft effective strategic communication to influence creation of value that's consistent with organizational objectives and performed with efficiency. Organizations may also find strategic communication useful in favorably influencing market perceptions and attitudes towards products and services in order for the business to achieve and accelerate market growth. Within an organization, at every point of use, well formulated strategic communication inspires and enables performance quality excellence, high efficiency, and improved value to customers because it clarifies the understanding of what needs to be accomplished in and through the organization. When strategic communication is poor, it serves as a potential obstruction in every phase of the productive effort thereby increasing and perhaps compounding the risk of poor performance. While talented employees and good processes can often counteract the effects of poor strategic communication the proper formulation and exercise of strategic communication can help employees and processes achieve their full potential.

In warfare, strategic communication is recognized as a significant asset. Opposing sides seek to continuously strengthen their strategic communication by integrating up-to-date guidance on coordinated efforts while they also seek to disrupt, confuse, and intercept enemy strategic communication in order to achieve competitive advantage. For example, with regard to successful naval operations against the Japanese during the Second World War, United States Army Chief of Staff George Marshall wrote:

> "...the Battle of the Coral Sea was based on deciphered messages and therefore our few ships were in the right place at the right time. Further, we were able to concentrate our limited forces to meet their naval advance on Midway when otherwise we almost certainly would

have been some 3000 miles out of place. We had full information on the strength of their forces..."[36]

George Marshall
1944

The Battle of Coral Sea demonstrates a principle regarding strategic communication that can be generalized for virtually all competitive endeavors.

The Coral Sea Principle

Excellent strategic communication improves performance quality and the strategic value of friendly forces actions while an adversaries' relatively poor strategic communication increases their risk of poor performance and contributes to their downfall.

Strategic communication corresponding to significant events in history is interesting to read because it brings to life the thoughts that shaped those moments in time. Looking back on historic strategic communication is also a fine way to introduce the concepts of strategy and quality by showing how events in history were influenced by the quality of strategic communication.

Considering the history of the United States, there are some very inspiring and enduring thoughts, particularly in respect to an individual's unalienable rights, which were effectively communicated in the U.S. Declaration of Independence.

"...We hold these truths to be self-evident, that all men are created equal, that they are endowed by their Creator with certain unalienable Rights, that among these are Life, Liberty and the

---

[36] Hearings Before the Joint Committee on the investigation of the Pearl Harbor Attack, Congress of the United States, pp. 1128-1129, https://archive.org/details/pearlharborattac03unit

pursuit of Happiness..."[37]

U.S. Declaration of Independence
July 4, 1776

In writing the Declaration of Independence, Thomas Jefferson sought to communicate an *acceptable* rationale for our original 13 colonies to unify as "We the people" and to pursue the Revolutionary War. This was done very *effectively* by invoking the "unalienable Rights" imparted to us by our "Creator" and then going on to list violations of these and other rights by the then governing King of Britain. It may well be said that beyond the essential and immediate purpose of declaring independence from Britain, the Declaration also set forth an appealing strategy to advance the colonists common quest for ascendant quality. This ascendant aspect of the Declaration's strategic communication was highlighted by our nation's 16th president, Abraham Lincoln.

> "They meant to set up a standard maxim for free society, which should be familiar to all, and revered by all; constantly looked to, constantly labored for, and even though never perfectly attained, constantly approximated, and thereby constantly spreading and deepening its influence, and augmenting the happiness and value of life to all people of all colors everywhere."[38]

Abraham Lincoln
1857

The message of the Declaration of Independence has resonated as a *sustaining* theme for our nation for more than two centuries in good times as well as in times which may otherwise have divided us. During the Civil War, President Lincoln reminded Americans of the Declaration of Independence, written four score and seven (87) years earlier, through the words of his Gettysburg Address.

---

[37] The Declaration of Independence, Thomas Jefferson, 6/28/1776, http://www.archives.gov/exhibits/charters/declaration_transcript.html

[38] Lincoln, Abraham. Speech on Dred Scott (1857), in Lincoln, The Collected Works of Abraham Lincoln, Vol. II, ed. Roy P. Basler (New Brunswick, NJ: Rutgers University Press, 1953), 403-407

"Four score and seven years ago our fathers brought forth, upon this continent, a new nation, conceived in liberty, and dedicated to the proposition that "all men are created equal."

Now we are engaged in a great civil war, testing whether that nation, or any nation so conceived, and so dedicated, can long endure.

...the nation, shall have a new birth of freedom, and... government of the people by the people for the people, shall not perish from the earth."[39]

Abraham Lincoln
November 19, 1863

The Declaration of Independence was also invoked by Martin Luther king, Jr. in his "I have a dream" speech, delivered from the steps of the Lincoln Memorial, as an *optimizing* theme encouraging our nation to ascend to a more ideal relationship among people.

"...I say to you today, my friends, that in spite of the difficulties and frustrations of the moment, I still have a dream. It is a dream deeply rooted in the American dream.

I have a dream that one day this nation will rise up and live out the true meaning of its creed: "We hold these truths to be self-evident: that all men are created equal."[40]

Martin Luther King, Jr.
August 28, 1963

The Declaration of Independence also expressed the ascendant idea that equal people have the unalienable right to *enrich the process* of living together as a society – through liberty and the pursuit of happiness. This ascendant ideal and prospect of a self-determined quality of life has been an effective message for those already in America and for those *throughout the world* seeking the beacon of

---

[39] Lincoln, Abraham. *Gettysburg Address* (Handwritten Draft), 11/19/1863; http://www.ourdocuments.gov/doc_large_image.php?doc=36
[40] King, Martin Luther. "*I have a Dream*," 8/28/1963. Retrieved on 5/31/2014; http://www.loc.gov/teachers/newsevents/events/creatingus/pdf/mlk.pdf

freedom. This message resonated with the nation of France where, through the effort of the sculptor Frédéric Bartholdi, the statue originally called "Liberty Enlightening the World" and now commonly known as the Statue of Liberty was created. The Statue of Liberty, a gift from the people of France, standing majestically on Liberty Island in New York harbor, is a sacred American image as she holds the beacon of freedom in one arm and the Declaration of Independence in the other. Unlike many statues that provide artistic impressions that are open to interpretation, the message being conveyed, in the revolutionary spirit of the Declaration of Independence, is unmistakable in the Statue of Liberty. A sonnet, by the poet Emma Lazarus, entitled "The New Colossus" is inscribed on a plaque at the pedestal of the Statue of Liberty where its message has welcomed countless immigrants to our country.

The New Colossus

"Not like the brazen giant of Greek fame,
With conquering limbs astride from land to land;
Here at our sea-washed, sunset gates shall stand
A mighty woman with a torch, whose flame
Is the imprisoned lightning, and her name
Mother of Exiles. From her beacon-hand
Glows world-wide welcome; her mild eyes command
The air-bridged harbor that twin cities frame.
"Keep, ancient lands, your storied pomp!" cries she
With silent lips. "Give me your tired, your poor,
Your huddled masses yearning to breathe free,
The wretched refuse of your teeming shore.
Send these, the homeless, tempest-tost to me,
I lift my lamp beside the golden door!"[41]

Emma Lazarus
1883

---

[41] Lazarus, Emma. "*The New Colossus*," 1883. Retrieved on 5/31/2014; http://www.libertystatepark.com/emma.htm

The message expressed in the Declaration of Independence has served our nation long and well and has proven itself to be:

- **Acceptable** as a vision unifying the 13 colonies as "We the people;"
- **Efficacious** in invoking our unalienable rights, and their violation, as a just cause for the Revolutionary War;
- **Sustaining** under the strain of civil war;
- **Optimizing** in times of imperfect peace;
- **Enriching** of our society's shared quest for liberty and the pursuit of happiness; and
- **World-class** in inspiring and inviting people in other nations who wish to live in freedom to join our nation's Quest for Ascendant Quality.

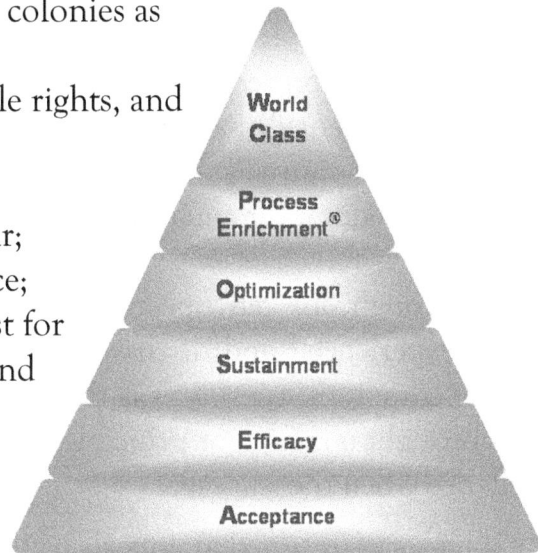

Figure 2-1: Quality's six Themes of Performance as expressed in The Declaration of Independence.

The impact on our culture and national identity made by Thomas Jefferson's approach to strategic communication was appreciated by journalist George Will.

"He gave us our creed. A lot of nations emerge from the mists of history and their basic identity is tribal, it's rooted in groups. Ours is rooted in a great ascent, an ascent to certain propositions. We are, as Lincoln said – Lincoln being the greatest student of Jefferson of them all – we are a nation dedicated to a "proposition."[42]

George Will
1997

In the Declaration of Independence we have a remarkable example of strategic communication that included, in a single sentence, a visionary strategy

---

[42] Burns, Ken. *"Thomas Jefferson - A Film By Ken Burns,"* 1997

that captured the identity and progressive aspirations of a people seeking to form a new nation. The Declaration also captured the strength and conviction in our nations' founders' collective will evident when war to remove the impediments to their progressive aspirations was preferable to the competing strategy of enduring taxation without representation in governance – a form of "conformance to the king's requirements" that our nation's founders saw as "absolute Despotism[43]" and "absolute tyranny[44]" – that had to be thrown overboard, like so many chests of tea, for freedom, for equality, and for the quest for ascendant quality.

Through our retrospective on the Declaration of Independence we have already experienced and discussed a rather sophisticated use of the elements of strategic communication. It was through this discussion that, with some subtlety, it was brought out that, in principle, strategic communication can achieve quality excellence when the message:

- ▶ expresses a strategy that appeals to the audience's quest for ascendant quality;
- ▶ expresses a current or future benefit over competing strategies; and
- ▶ expresses the desired level of quality in each of quality's six Themes of Performance such that quality excellence is, to borrow President Lincoln's phrase, "familiar to all, and revered by all; constantly looked to, constantly labored for, and even though never perfectly attained, constantly approximated, and thereby constantly spreading and deepening its influence, and augmenting" the value of performance quality everywhere.

These first two points are, by now, ideas that we can be comfortable with. On any issue of concern, we appreciate strategic communication in which we can find solutions to our challenges that operate to improve our quality of life. Our third point bears an introduction to a new concept – quality's six Themes of Performance.

In essence, quality's six Themes of Performance can be thought of as six unique, hierarchical, themes - central concepts, impressions, or viewpoints – that, taken together, are capable of comprehensively describing the performance quality of any thing or of any idea. Quality's six Themes of Performance are

---

[43] Jefferson, Thomas. The Declaration of Independence
[44] The Declaration of Independence

pictured, in Figure 2-1, along with a brief qualitative assessment, by Theme of Performance, of the performance quality of the Declaration of Independence over the course of United States history. As you may now appreciate, having read through this qualitative assessment, with little prior preparation, when the Themes of Performance are used, it's easy to communicate the quality of something as complex as the Declaration of Independence. What's more, since the thought process in defining quality through the Themes of Performance is consistent with the way people routinely conceptualize quality, intuitively, all day long, the Themes of Performance enable you to better and more correctly express what you're thinking. This correspondence between your intuitive methods for assessing quality and quality's Themes of Performance can become more apparent as we review the assessment in Figure 2-1 more closely.

▶ In the **Acceptance** <u>Theme of Performance</u> we find that the Declaration was "Acceptable as a vision unifying the 13 colonies as "We the people." This acceptability of the Declaration was critical because if the Declaration was *unacceptable*, colonists would undoubtedly demand a new Declaration written to describe ideals more closely aligned with a different vision.

▶ In the **Efficacy** <u>Theme of Performance</u> we build on an acceptable Declaration and see that the Declaration was also "Efficacious in invoking our unalienable rights, and their violation, as a just cause for the Revolutionary War." Although the Revolutionary War was already taking place at the time of the signing of the Declaration, the fact that the Revolutionary War continued, under the moral force and for the stated rationale in the Declaration, is strong evidence that the Declaration was efficacious in support of the ongoing war.

▶ In the **Sustainment** <u>Theme of Performance</u> we can see that, in the long term, the acceptable and effective quality of the Declaration was "Sustaining under the strain of civil war" as it served President Lincoln, when he referred to it, in the midst of the Civil War, in his Gettysburg Address.

▶ In the **Optimization** <u>Theme of Performance</u> we can see that the acceptable, efficacious, and sustained value of the Declaration could be recurred to as the moral justification Dr. King drew upon to help improve equality among people. This showed the Declaration to be "Optimizing in times of imperfect peace."

- ▶ In the **Process Enrichment** Theme of Performance, which is concerned with motivation of people in the system and their working together towards Coascendancy, the Declaration empowered people to make laws for themselves in a representative form of government which was "Enriching of our society's shared quest for liberty and the pursuit of happiness."

- ▶ In the **World-class** Theme of Performance, the Declaration, having achieved quality excellence in all the lower level Themes of Performance, and also being consistent with the aspirations of people in other parts of the world, was "World-class in inspiring and inviting people in other nations, who wish to live in freedom, to join our nation's Quest for Ascendant Quality."

Creating quality excellence in strategic communication – or in anything else for that matter – requires creating quality excellence in each of quality's six Themes of Performance. As we will soon see, deviation from quality excellence in any Theme of Performance is a risk of poor performance. For this reason, we will closely study the influence and effect of deviation from quality excellence in each Theme of Performance so that we can recognize and properly respond to changes in the balance of quality and risk that run contrary to our intended strategy.

Quality's six Themes of Performance are central to a new definition of quality provided by the Process Enrichment philosophy.

QUALITY

Quality is the ability of performance, in each Theme of Performance, to enact a strategy.

Based on the Process Enrichment philosophy's definition of quality, we can appreciate that the purpose of the concept of quality is to serve as an indicator of the ability of performance of some thing or of some idea to attain an incentive we value through a strategy we have envisioned.

Since quality is a concept that's linked to performance, and everything performs in some way – even if you have to really think for a while to figure out what something does – quality is not reserved only for the things we can hold in our hands or that we can sense in any other way, rather, everything possesses some form of quality because of its ability to perform.

Even Leonardo da Vinci's Mona Lisa, in which the portrait appears to be rather relaxed, is performing the function of activating your imagination and creating a state of mind that's desirable. The Mona Lisa's excellent ability to create a desirable state of mind, which is the primary incentive derived from the painting, is why the painting is said to have quality excellence and extraordinary value. Quality isn't always about the performance a product or service does for you – sometimes quality is about the performance done to you – but quality is always about performance.

Figure 2-2;
Leonardo da Vinci's
Mona Lisa, 1519

Quality is valued by us because the better the quality in some thing or in some idea, the better that thing or idea can help us perform a strategy which we believe will support or advance our quest for ascendant quality. This is the basic logic of the concept of quality.

Strategy is valued by us because it is through the performance of a strategy that we preserve and improve our well-being through attainment of the incentives we desire. To attain incentives like money, food, shelter, or various forms of enjoyment, we must take chances on a strategy that we devise which we believe is within our capabilities and in which we reasonably expect success. Since quality is best when it indicates performance that can make a strategy succeed, intentions for quality, necessarily, follow from an understanding of the intentions of a strategy. Our strategy, then, is the reference standard against which the ability to perform of our acquisitions or of our creative efforts should be compared. For this reason, the definition of strategy that we have informally cultivated thus far should be formally defined in a manner in which we can find an improved ability to formulate strategy and acquire or create quality excellence that helps us succeed in attaining our desired incentives. A new definition of strategy that can serve these purposes is provided by The Process Enrichment philosophy.

The Process Enrichment philosophy's definition of strategy is intended to be consistent with the natural flow of thought and the essential elements of strategy that people actually consider when they devise a strategy that guides their behavior. Strategies developed for military warfare, competitive warfare, criminal enterprises, surgical procedures, and for the peaceful pursuit of the quest for ascendant quality are simply the same conceptual system used for a variety of different purposes.

The defining example of strategy is George Washington's strategy of warfare which he developed as Commander-in-Chief of the Continental Army during the American Revolutionary War in 1776. Mr. Washington formulated a strategy of war which expressly recognized that quality in his war strategy required the avoidance of unnecessary risk to revolutionary America's troops – upon which everything depended.[45] Mr. Washington wrote of his strategy for the Revolutionary War in a letter to the Continental Congress.

> "...on our Side the War should be defensive. It has even been called a War of Posts. That we should on all Occasions avoid a general Action, or put anything to the Risque, unless compelled by a necessity, into which we ought never to be drawn.

---

[45] Revolutionary troops of the American colonies were what some modern-day military strategists would call a "Center of Gravity" based on the ideas of Prussian General Carl von Clausewitz in his 1832 book "On War." The concept of Center of Gravity is explored in greater detail in chapter 3.

The Arguments on which such a System was founded were deemed unanswerable and experience has given her sanction. With these views, and being fully persuaded that it would be presumption to draw out our Young Troops into open ground, against their Superiors both in number and Discipline; I have never spared the Spade and Pick Ax; I confess I have not found that readiness to defend even strong Posts, at all hazards, which is necessary to derive the greatest benefit from them. The honor of making a brave defence does not seem to be a sufficient stimulus, when the success is very doubtful, and the falling into the Enemy's hands probable."[46]

George Washington
1776

In Mr. Washington's expression of the Colonial Army's defensive strategy, we can find directly or by inference that all the elements of a <u>conceptual "System"</u> of strategy are duly represented. We can find the <u>incentive</u> to preserve troops by forgoing the "honor of making a brave defence" which makes their "falling into the Enemy's hands probable;" we can find <u>expectations regarding the incentive</u> in the expected use of the preserved troops towards eventual victory – an expectation validated by the rationale of "unanswerable" arguments and "experience;" we can find the <u>motive expressed as an intention</u> that "the War should be defensive;" and we can find that the intentions <u>are carried out through behaviors using resources</u>, where behaviors include "on all Occasions avoid a general Action" and avoidance of the behavior of drawing out "Young Troops into open ground" - where troops can be considered resources. Mr. Washington's strategy also expresses the required <u>degree of quality</u> as the avoidance of actions that "put anything to the Risque, unless compelled by a necessity, into which we ought never to be drawn." Mr. Washington's concise expression of his strategy for the Continental Army is a remarkable reflection of his deep understanding of the elements of strategy and quality and, also, of the importance of strategic communication with the people whose support was

---

[46] George Washington, Letter to the Continental Congress, 9/8/1776, http://www.loc.gov/teachers/classroommaterials/presentationsandactivities/presentations/timeline/amrev/north/posts.html

essential. We can also appreciate that Mr. Washington's strategy governed "the first successful revolution against a colonial empire in world history."[47]

The principal reason for emulating the thoughts and behaviors of people in the definition of the conceptual system of strategy is to benefit from the inherent skill and experience people have exhibited in developing and executing strategy, over the course of thousands of years, to preserve and improve their well-being. Consider, for example, the ease with which each of us develops strategy to satisfy our basic need for food. If we are hungry, food is the incentive of our strategy and we go food shopping or, rather than buying food on impulse when we're already hungry, we earn and save enough money to pay for the healthy food we want and we shop in advance of our need for food. As our strategy to meet our need for food improves, the quality of our meal improves and our risk of a poor meal declines proportionately. Of course, in developing our strategy, we establish the desired level of quality in consideration of how much we are ready, willing, and able to pay to attain it as a means of balancing the alternatives open to us. This is accomplished through a simultaneous balancing of our competing needs for our limited financial and other resources. For example, even though we may, ideally, enjoy a fillet mignon dinner, quality excellence, considering a smaller budget, may be achieved with a hamburger and French fries. Also, while the acceptable requirement for a meal is that we have food, a fancy, full service, restaurant may better address our desires in the higher level Themes of Performance than the take out window at a fast food restaurant – when we can take the time to enjoy it. This is the basic system in which strategy, quality, and risk are interrelated. It is because of our deeply entrenched natural talent and experience, using strategy to meet our personal needs, that we can and should apply this talent to developing our business strategies or even to developing our war strategies.

The discovery of quality's six Themes of Performance and the establishment of the definitions of quality and strategy, presented here for the first time, were the products of several years of research and creative effort. An important method used to test the logic and validity of these concepts was the analysis of various forms and grades of strategic communication. Strategic

---

[47] Wikipedia / Cunliffe, Marcus (1958). George Washington, Man and Monument. Boston: Little, Brown. OCLC 58007859.

communication has a benefit in testing these concepts because historians have described events leading up to and following historic strategic communication from which the effects of strategic communication can often be clearly discerned. For this reason, our study of the Declaration of Independence and George Washington's war strategy has initiated us to the method through which we can continue to validate the new concepts and definitions introduced by the Process Enrichment philosophy and attain confidence in their implications for our creative effort.

Victories of the human spirit, in various forms of endeavor, are not always accompanied by glorious words – and strategic communication isn't always as inspiring as it could be – even when delivered by a recognized leader. This doesn't, necessarily, mean that the underlying purpose is wrong or that the leader's principles are wrong, it's a matter of proper delivery of the proper elements of strategic communication with a proper message. For example, British Prime Minister Neville Chamberlain's radio broadcast of Britain's declaration of war against Germany, which initiated Britain's military involvement in the war that came to be known as World War II (WWII), may not have been considered very inspiring. An excerpt of this speech shows its flavor.

"I am speaking to you from the Cabinet Room at 10, Downing Street.

This morning the British Ambassador in Berlin handed the German Government a final note stating that, unless we heard from them by 11 o'clock, that they were prepared, at once, to withdraw their troops from Poland, a state of war would exist between us.

I have to tell you, now, that no such undertaking has been received, and that consequently this country is at war with Germany.

You can imagine what a bitter blow it is to me that all my long struggle to win peace has failed. Yet I cannot believe that there is anything more or anything different that I could have done and that would have been more successful...”[48]

Neville Chamberlain
September 3, 1939, 11:15 AM

This speech certainly ended Mr. Chamberlains’ strategy of, what has been called, “appeasement” of Germany, and provided some defensive justification for having maintained it, but the speech offered little in the way of a forward looking strategy for prosecuting the war. In terms of quality, because it met its essential requirement of declaring that Britain was at war with Germany, the speech met expectations in quality’s Acceptance Theme of Performance, however, considering quality’s higher echelon Themes of Performance, from Efficacy on up, the absence of an inspirational course of action in the speech served rather poorly.

Ironically, Mr. Chamberlain’s political rival, Winston Churchill, who effectively used strategic communication to increase the risk of poor performance of Mr. Chamberlain’s policy of appeasement, through various speeches and written newspaper articles in opposition to it, became Mr. Chamberlain’s successor and he also became one of the most influential leaders and one of the most famous speakers of the last century. This isn’t necessarily an indication that Mr. Churchill served better purposes or principles than Mr. Chamberlain – it simply recognizes that Mr. Churchill had an exemplary talent in strategic communication that could be used for either right or wrong aims.

One of Mr. Churchill’s most famous speeches was presented in Britain’s House of Commons on the day following the “Miracle of Dunkirk.” The Miracle of Dunkirk was the name given to an event near the beginning of WWII, when nearly 200,000 British and well over 100,000 French troops survived a week-long retreat, under heavy enemy fire, through the port of Dunkirk in northern France. Hundreds of small boats brought the retreating soldiers across the English Channel to safety in Britain. At the time of this speech, in view of the retreat and the battle loss of men and equipment in France, and in view of the evident lack

---

[48] Neville Chamberlain, broadcast of Sept. 3, 1939 at 11:15 AM; http://www.bbc.co.uk/archive/ww2outbreak/7917.shtml

of preparation for modern warfare, it was uncertain whether, or for how long, Britain would continue in the war. With the ears of the world, of friends and foes alike, pressing towards their radios, for the news that would influence the fate of many people and many nations, Mr. Churchill delivered his historic speech.

We Shall Fight on the Beaches

"... we shall prove ourselves once more able to defend our Island home.

... We shall go on to the end, we shall fight in France, we shall fight on the seas and oceans, we shall fight with growing confidence and growing strength in the air, we shall defend our Island, whatever the cost may be, we shall fight on the beaches, we shall fight on the landing grounds, we shall fight in the fields and in the streets, we shall fight in the hills; we shall never surrender, and if, which I do not for a moment believe, this Island or a large part of it were subjugated and starving, then our Empire beyond the seas, armed and guarded by the British Fleet, would carry on the struggle, until, in God's good time, the New World, with all its power and might, steps forth to the rescue and the liberation of the old."[49]

Winston Churchill
June 4, 1940

This is a carefully crafted message communicating to the world Britain's forward looking intentions. There are several elements of this speech that make it stand out as effective strategic communication. It sounds good to the ear, it is emotionally engaging and, while acknowledging the risks, it succinctly expresses Britain's war strategy with a success oriented, ascendant, quality that served to align people to a shared vision. Also, as it turned out, the speech was prophetic in anticipating the air battle that Mr. Churchill branded the Battle of Britain, the sea battle that Mr. Churchill branded the Battle of the Atlantic, and the eventual participation of the New World in the war.

---

[49] Winston Churchill, radio broadcast, 6/4/1940, Sound recordings, as well as the full text, of this speech are available on the Web. http://www.youtube.com/watch?v=MkTw3_PmKtc

Mr. Churchill's "We Shall Fight on the Beaches" speech encompasses the elements of strategic communication in a manner that bears close study. My research has shown that, as unique as this speech is, the elements that make this speech excellent strategic communication aren't unique to this speech. In fact, in a study of dozens of historically significant examples of excellent strategic communication, dating as far back as the Ten Commandments and also including the several writings and speeches quoted in this chapter, these elements were found with great regularity. This realization, as well as the realization that deficiencies in these elements are a causal factor in poor strategic communication and, in some cases, tragic strategic outcomes, spurred development of the Ascendant Strategy Model.

The Ascendant Strategy Model

The crafting of strategic communication that's intended to influence thought and behavior, inspire revolutions, secure and perfect equality or help add net income to the bottom line, is often thought of as an artistic talent. We all know people who could be expected to communicate certain types of information more effectively, more captivatingly or more on point. That's probably why the founders of our country sought out Thomas Jefferson to write the Declaration of Independence and that's probably why oceans of people turned out to hear Martin Luther King Jr. speak on several occasions. Fortunately, for those of us who seek to master strategy and strategic communication and the creation of quality excellence, there are interrelationships and patterns which, like knowledge of music theory or geometry, serve to elevate your personal talent to the level of a skilled artist. The pattern of strategy is one such pattern and the hierarchal pattern of quality's six Themes of Performance is another, and they both work together in the intuitive context of the Ascendant Strategy Model.

A study of Mr. Churchill's "We Shall Fight on the Beaches" speech, in the context of the Ascendant Strategy Model will allow us to gain a deeper understanding of why this speech was effective as we also gain useful insight to the elements and structure of strategic communication. The pattern of Mr. Churchill's high level war strategy, as expressed in his "We Shall Fight on the Beaches" speech, is shown, in Figure 2-3, as it corresponds with the Ascendant

Strategy Model. (Terminology from military strategy in the Ends – Ways – Means construct is shown in green text and terminology from business strategic planning construct is shown in black text as references for those already familiar with them. The Ascendant Strategy construct is shown in orange text.)

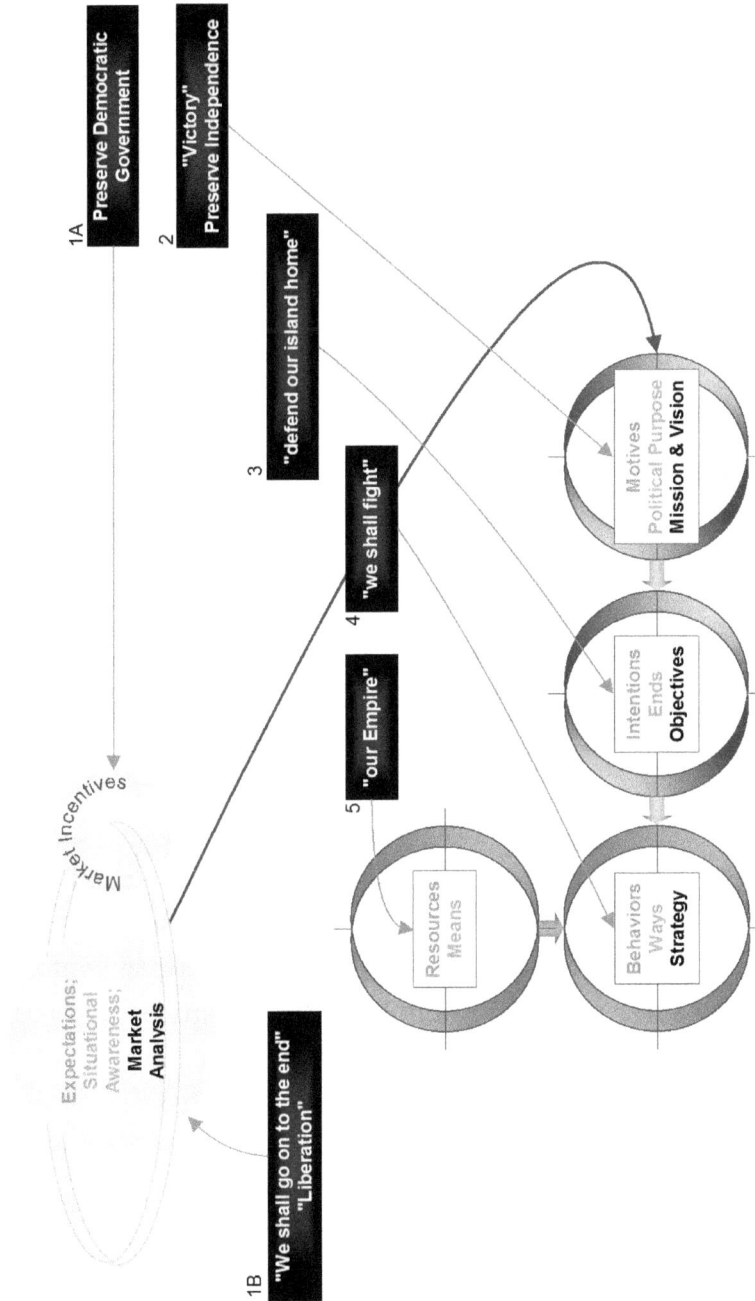

Churchill's Strategy: Market Incentives (Preserve Democratic Government, 1A) and expectations regarding them ("We shall go on to the end", "Liberation," 1B) create motives ("Victory," Preserve Independence, 2) expressed as intentions ("We shall defend our island home," 3) which are carried out through behaviors ("we shall fight," 4) using resources ("Our Empire," 5).

Figure 2–3; The Ascendant Strategy Model (part 1) showing Mr. Churchill's top level strategy in WWII as expressed in his "We Shall Fight on the Beaches" speech.

Stepping through Figure 2-3, we can recognize, based on his "We Shall Fight on the Beaches" speech, that one of Mr. Churchill's <u>incentives</u> was to preserve Britain's democratic government (1A) rather than submit, under pressure of war, to Nazism. We can also recognize that his <u>expectations</u> with regard to preserving Britain's parliamentary democracy were that Britain, and also the countries endangered by Nazi aggression, would go on to the end of the war and successfully achieve liberation (1B) – retaining their desired forms of government. This established the <u>motive</u> in Mr. Churchill's strategy – to achieve military victory (2) over the Nazis and thereby preserve independence. Mr. Churchill's planned <u>intention</u> was to "defend our island home" (3). The <u>behavior</u>, or way to perform defense, was expressed as "we shall fight" (4) and the <u>resources</u> to be used are those in, as he said, "our empire" (5) and eventually those of the New World.

In Mr. Churchill's "We Shall Fight on the Beaches" speech, we find the product of a skilled speaker and leader, who was evidently very aware of the conditions and circumstances of the war situation and of the course of action to which he wished to lead his nation – as well as to the state of mind that he wished the audience for this speech to attain. It was this strategic communication that bolstered the French, who were still fighting, reached out to America, and set Britain to the task of building organizations to perform the necessary defensive and offensive activities and to also produce the products and services necessary to win the war.

Another remarkable observation about Mr. Churchill's "We Shall Fight on the Beaches" speech is that, in addition to expressing the key elements of strategy, the concluding paragraph of Mr. Churchill's speech also expressed, at a high level, the performance necessary, in each Theme of Performance, to prosecute Britain's war effort. As shown in figure 2-4, the progression of Mr. Churchill's speech followed the same order as the classic pattern of quality's six Themes of Performance.

Intentions of Mr. Churchill's War Strategy
In Each of Quality's Six Themes of Performance
As Expressed in his "We Shall Fight on the Beaches" Speech

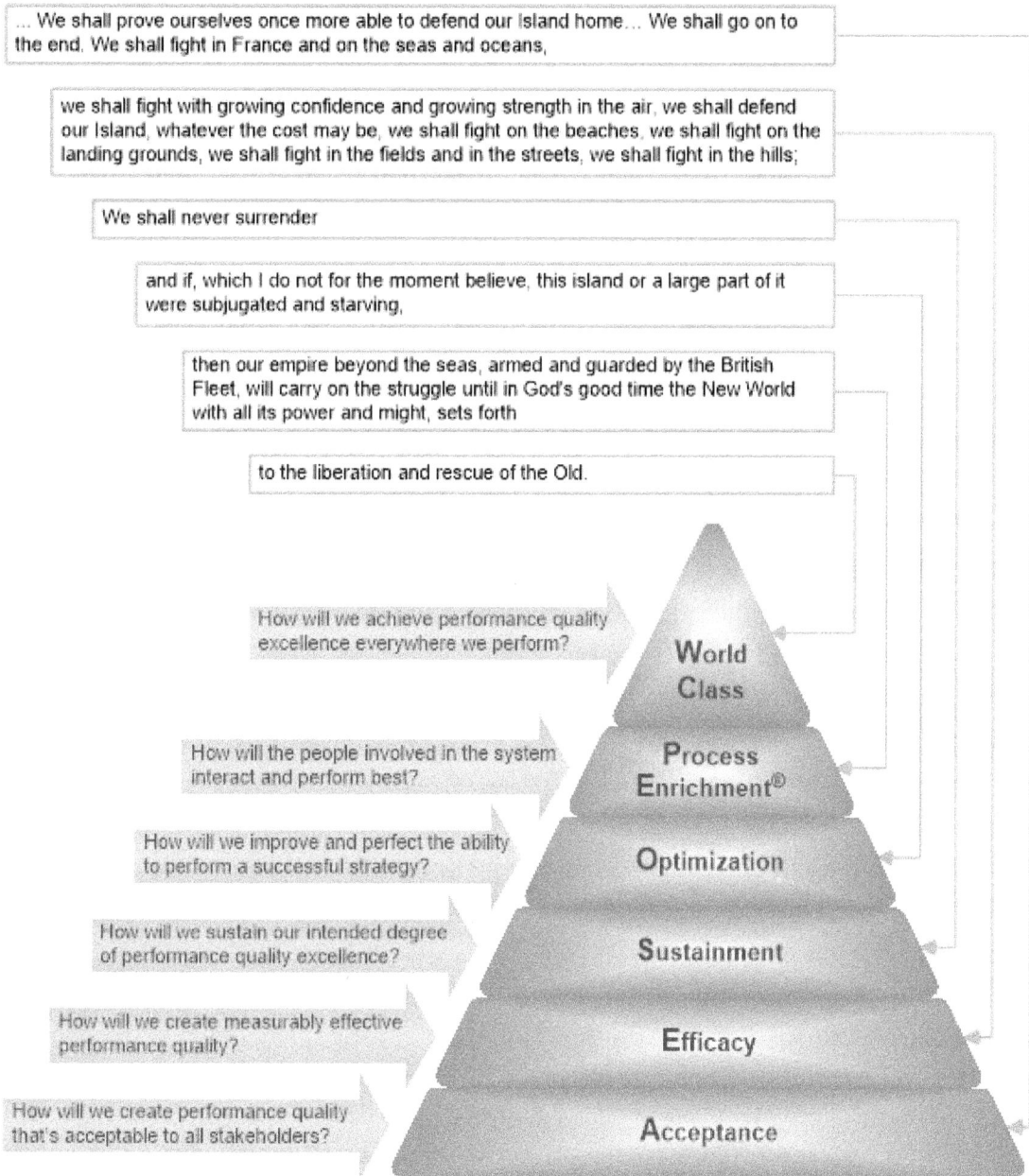

... We shall prove ourselves once more able to defend our Island home... We shall go on to the end. We shall fight in France and on the seas and oceans,

we shall fight with growing confidence and growing strength in the air, we shall defend our Island, whatever the cost may be, we shall fight on the beaches, we shall fight on the landing grounds, we shall fight in the fields and in the streets, we shall fight in the hills;

We shall never surrender

and if, which I do not for the moment believe, this island or a large part of it were subjugated and starving,

then our empire beyond the seas, armed and guarded by the British Fleet, will carry on the struggle until in God's good time the New World with all its power and might, sets forth

to the liberation and rescue of the Old.

How will we achieve performance quality excellence everywhere we perform?
World Class

How will the people involved in the system interact and perform best?
Process Enrichment®

How will we improve and perfect the ability to perform a successful strategy?
Optimization

How will we sustain our intended degree of performance quality excellence?
Sustainment

How will we create measurably effective performance quality?
Efficacy

How will we create performance quality that's acceptable to all stakeholders?
Acceptance

Figure 2-4, Intentions of Mr. Churchill's war strategy, in each of quality's six Themes of Performance, as expressed in his "We Shall Fight on the Beaches" speech, and generic focusing questions for each Theme of Performance.

The Acceptance Theme of Performance

How will we create performance quality that's acceptable to all stakeholders?

In Mr. Churchill's strategic thinking, the incentive to preserve democratic government and the expectation that Britain could "go on to the end" created the motive to achieve victory in the war. On this strategic basis, Mr. Churchill, at a minimum, was compelled to communicate the intentions to continue the war and to achieve victory. A speech on continuing the war that didn't propose winning would probably be unacceptable to the relevant stakeholders. In this instance, the stakeholders were the people who lived in Britain, the people who had to fight the war, the people in the rest of Britain's empire – who would also be affected by the decision to continue the war, and also allied nations and nations threatened by Germany in the event that Britain decided not to fight. Having decided upon continuing in the war, Britain would, necessarily, fight everywhere that Germany was capable of fighting – on the land, sea, and air. It was therefore best – although not essential – to express the intention to fight on the land, sea, and air in order to spur the planning necessary to organize for building the products and services necessary to achieve victory.

Mr. Churchill's intentions in the Acceptance Theme of Performance were expressed as follows:

> "... We shall prove ourselves once more able to defend our Island home... We shall go on to the end, we shall fight in France, we shall fight on the seas and oceans..."

To accomplish these top level intentions, in the Acceptance Theme of Performance, we can use the six Themes of Performance, once again, to identify appropriate subordinate level intentions which may be easier to delegate and manage. For example, a reasonable set of subordinate level intentions (enacted as behaviors) could be the following:

Acceptance:
- The armed forces shall be capable of operations in France and on the oceans

Efficacy:

- The armed forces will be equipped to fight as necessary

Sustainment:
- Supply and re-supply shall be effectuated

Optimization:
- Compose forces appropriately for battles

Process Enrichment:
- Apply strategic communication to motivate resistance

World Class Performance:
- Resist and defend everywhere

This ability to use the Themes of Performance recursively is a significant and unique advantage within our definition of quality because it allows us to decompose to an infinite level of detail the performance quality necessary of intentions, subordinate intentions, sub-sub-intentions, and so on. In this way, throughout an organization, significant aspects of organizational strategy can flow down to operational and product and service intentions with very low risk that key intentions and necessary performance will be overlooked or misunderstood. Another way of stating this is that, due to the Themes of Performance, our definition of quality has breadth of coverage of quality needs and due to the ability to use the definition of quality recursively, throughout an organization, we also have depth of coverage of quality needs. Because of these properties of breadth and depth, and generic applicability to any creative effort, our definition of quality enables an organization to make the best use of their strategy and strategic communication, throughout the organization, utilizing a common language, methods, and philosophy.

The recursive use of the Themes of Performance is an advanced application – with the most common application being in the construction of a Balanced Enterprise Strategic Themes (BEST) Framework. The BEST Framework is a tool for developing and organizing strategy and creating strategic communication. A BEST Framework modeling Britain's war strategy, as expressed in Mr. Churchill's "We Shall fight on the Beaches" speech, is presented in Figure 2-5. The details of our continuing discussion of each of the six Themes of Performance can be followed using the BEST Framework pictured in Figure 2-5. For example, the Acceptance Theme intentions and the decomposition to behaviors, just discussed, are shown in Figure 2-5 in the "Acceptance" Theme column. Note that since intentions describe planned behaviors the description of behaviors and intentions are roughly the same.

The Efficacy Theme of Performance

How will we create measurably effective performance quality?

In Mr. Churchill's speech, having indicated that war would be pursued, to be effective, industry needed to be mobilized to build more of the necessary equipment; people needed to know that the effort was funded; and they needed to know the ways they may be called upon to assist in the war effort. In the Efficacy Theme of Performance Mr. Churchill said:

> "we shall fight with growing confidence and growing strength in the air, we shall defend our Island, whatever the cost may be, we shall fight on the beaches, we shall fight on the landing grounds, we shall fight in the fields and in the streets, we shall fight in the hills . . . ."

As with all of the Themes of Performance, simply considering the Theme, and what's necessary to achieve quality excellence in it, makes the type of measurement, appropriate to assessing performance, almost jump out at you. Here are some basic considerations for developing Thematic Performance Measurements:

▶ Measurements in The Acceptance Theme assure that essential stakeholder needs for the creative effort are covered (fully addressed).

▶ Measurements in The Efficacy Theme assure that the creative effort's performance is measurably effective.

▶ Measurements in the Sustainment Theme assure that the creative effort's operational performance capabilities are maintained.

▶ Measurements in the Optimization Theme assure that the creative effort's design-best implementation is achieved.

- Measurements in the Process Enrichment® Theme assure that the creative effort works for the people in the system.
- Measurements in the World-class Theme of Performance assure that the creative effort is valued wherever it is used.

As we can see, in the Efficacy Theme, Mr. Churchill's emphasis on growing confidence and growing strength in the air means a progressive effort that can be measured, over time, for example, by the growth in the number and types of planes and growth in the number of trained and effective pilots that can, and do, win air battles – thereby growing confidence in the air.

Based on Mr. Churchill's intention in the Efficacy Theme of Performance, it's possible to conceive of behaviors to accomplish this intention by, once again, applying the definition of quality to ensure that the intention, to be effective on the land, sea, and air, is performed with quality excellence. Subordinate intentions and behaviors, by Theme of Performance, could be:

Acceptance:
- All zones of operation will be defended

Efficacy:
- Add/improve weaponry & soldiers in needed competencies

Sustainment:
- Integrate Lessons Learned into plans and processes

Optimization:
- Design & Build specialized weaponry

Process Enrichment:
- Destroy/disrupt enemy capabilities & enemy will to fight

World Class Performance:
- Win all battles

The Sustainment Theme of Performance

How will we sustain our intended degree of performance quality excellence?

In Mr. Churchill's speech, performance quality excellence in sustainment of the war effort was stated simply and unequivocally when Mr. Churchill said:

"We shall never surrender"

The intention not to surrender could be supported and sustained by the following subordinate intentions and behaviors:

Acceptance:
- No negotiation with enemy combatants (on terms of surrender)

Efficacy:
- Dominate the enemy's message; communicate about successes

Sustainment:
- Protect supplies

Optimization:
- Allies commit resources to vision of victory

Process Enrichment:
- Communicate the rationale for not surrendering; rationale for new resistance

World Class Performance:
- Win the war; sustain the effort to win everywhere

The Optimization Theme of Performance

How will we improve and perfect the ability to perform a successful strategy?

It's important to realize that Britain is an island that's dependent on shipping in order to import an adequate supply of food – which isn't available on the island. Since Britain's shipping would be endangered during the war it would be optimal to successfully avoid starvation and perform behaviors to avoid that circumstance and carry on necessary actions even under difficult conditions. In this context, in Mr. Churchill's speech, performance quality excellence in optimization was expressed as:

"and if, which I do not for the moment believe, this island or a large part of it were subjugated and starving,"

The intention to optimize for the anticipated loss of imports during conditions of war could be accomplished through the following subordinate intentions and behaviors:

Acceptance:
- Control resources

Efficacy:
- Allocate resources as needed; find new sources

Sustainment:
- Stretch resources; use alternatives

Optimization:
- Distribute resources efficiently by priority

Process Enrichment:
- Employ rationing system for rare resources

World Class Performance:
- Survive

The Process Enrichment® Theme of Performance

How will the people involved in the system interact and perform best?

In Mr. Churchill's speech, performance quality excellence in the Process Enrichment Theme of Performance means that all the people currently involved in the war effort and those who, perhaps will be, will work together and share resources to meet strategic intentions. This was expressed as:

> "then our empire beyond the seas, armed and guarded by the British Fleet, will carry on the struggle until in God's good time the New World with all its power and might, sets forth"

The intention to work with the people in the system could be advanced by the following subordinate intentions and behaviors:

Acceptance:
- Preserve the fleet to enable military assistance from allies

Efficacy:
- Gain active participation of allies in the war effort

Sustainment:
- Use strategic communication to promote mutual aspirations

Optimization:
- Integrate shared resources

Process Enrichment:
- Worldwide allied cooperation and aid

World Class Performance:
- Never give in

The World Class Theme of Performance

How will we achieve performance quality excellence everywhere we perform?

In Britain's war effort, it was important that this major undertaking, and hopefully victorious result, was a shared victory amongst the Old World countries that were overrun by Germany – so that complete success could be achieved wherever the war raged. This was expressed in Mr. Churchill's speech as:

"to the liberation and rescue of the Old."

The intention to prosecute the war effectively, and wherever necessary, could be supported and sustained by the following subordinate intentions and behaviors:

Acceptance:
- Achieve liberation

Efficacy:
- Gain unconditional surrender of all enemy combatant nations

Sustainment:
- Work with allies towards common goals

Optimization:
- Reduce time to liberation

Process Enrichment:
- Eliminate enemy influence

World Class Performance:
- Liberate all territory, restore overthrown governments

The BEST Framework

The BEST framework has its origins in the work popularized by Art Schneiderman, and later, Robert Kaplan and David Norton, in the Balanced Scorecard – which has grown in use as a vehicle for strategic communication. Both the BEST Framework and the Balanced Scorecard support the strategic planning process and produce an at-a-glance, high level, view of organizational strategy.

The BEST framework was developed simultaneously with the Process Enrichment philosophy, which was not available at the time the Balanced Scorecard was designed. For this reason, the BEST Framework is able to benefit from the newly defined concepts of strategy, quality, and risk and their interrelationship as described in the Ascendant Strategy Model. As shown in Figure 2-5, The BEST Framework is organized from top to bottom with the elements defining strategy in their intuitively proper sequence. Along the left edge of the BEST chart, you will find references to elements of the Process Enrichment oriented elements of strategy. Along the far right edge are the traditional elements used to represent strategy in the Department of Defense's Ends-Ways-Means construct as described in Joint Publications JP-3 through JP-5. The integrated use of quality's six Themes of Performance is also evident in a progression from left to right of six top level "intentions" or "objectives," corresponding to each Theme of Performance. Each top level intention is, itself, further resolved, once again, into the six Themes of Performance vertically to create the 36 element lattice of component strategic intentions. Each of the 36

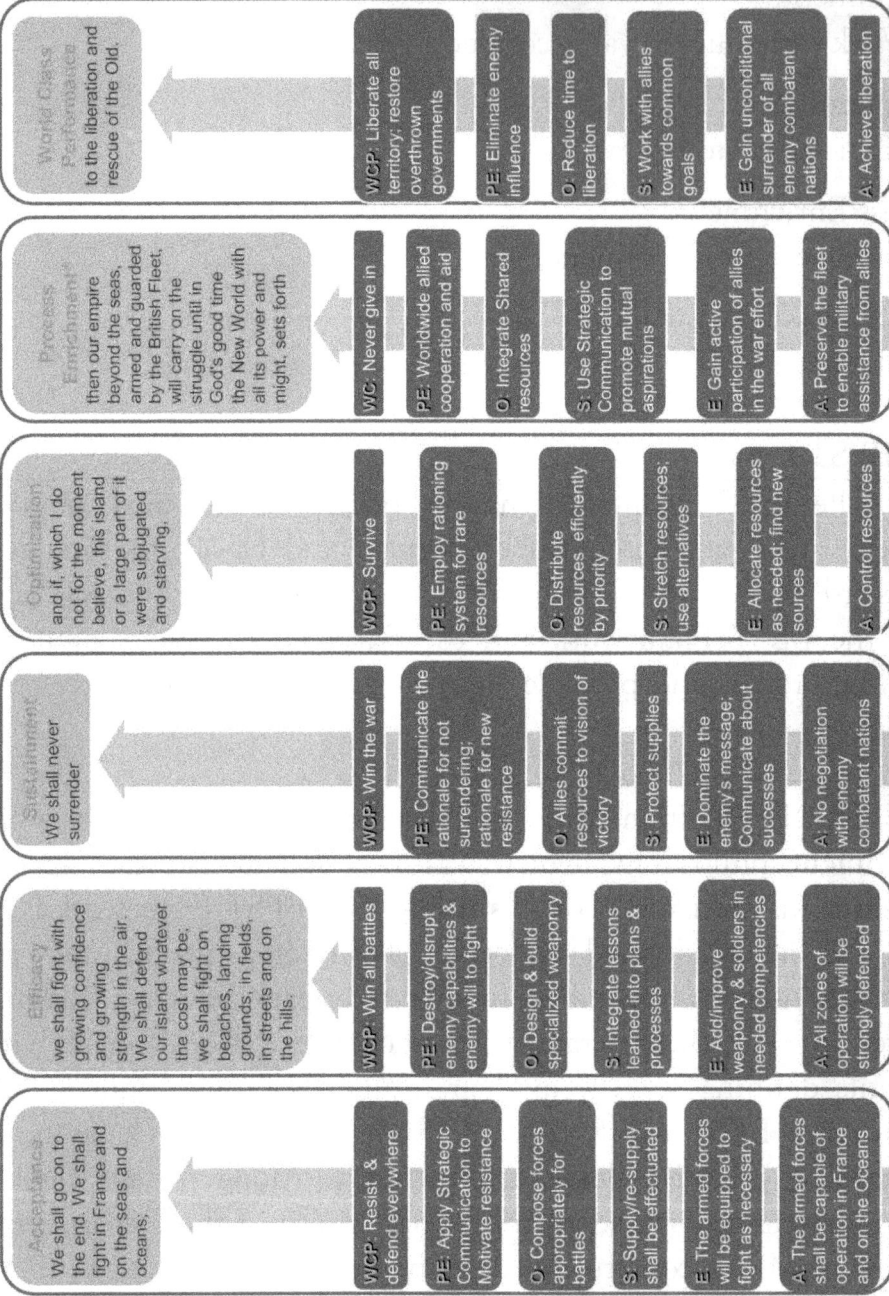

Intentions (or behaviors) in the lattice can be broken out to separate BEST framework charts where they can be further decomposed into desirable intentions or goals. In this way, separate functional groups or business units can manage strategic performance at an appropriate level of detail. The clarity and thoroughness of BEST charts serves to ensure that necessary intentions or goals are identified and, subsequently, accomplished. BEST charts can also be used to ensure that performance is managed and reported, as necessary, throughout an organization.

The BEST Framework serves the strategic planning process by helping to ensure Content Coverage for top level organizational objectives - making sure that they are properly elicited from stakeholders and that they comprehensively cover the organization's needs. In this way, the BEST framework provides the basis for the organization's Strategic Communication. The BEST Framework should be produced as the culminating part of market analysis, in the context of organizational capabilities, establishing forward looking planned intentions to compete in the marketplace with the organizations products and services. The BEST Framework, although it may appear complex, is no more complex than is necessary to ensure that important objectives aren't left out. In the same way that criteria or checklists can be effective in ensuring the success of a process, the BEST framework can enable strategists to simultaneously think about incentives in the market, process performance and resource requirements necessary to attain them, and quality considerations to assure effectiveness in competition.

A key advantage of the BEST Framework's structure is that it draws your focus to 36 unique intentions individually rather than simultaneously - in this way the specification of intentions is comprehensive, is of high quality, and is relatively easy to create in a step-by-step manner. Since it is customary in an organization to peer review and revise intentions over a series of meetings this step-by-step process becomes a great advantage over brainstorming because the management team can easily pick up where they left off. The intentions produced also have traceability to higher level intentions - a property that is inherent in the BEST Framework's method of construction. This removes the inconsistency and relevance issues that can corrupt organizational intentions and objectives when unstructured and undisciplined techniques such as brainstorming are used to specify them. The intentions specified through the

BEST framework can also serve as the foundation for quantifying business strategy through an organizational measurement system that enables management to track the performance of organizational strategy.

Since our definition of quality establishes a common language within an organization, the flow of information can also happen in both directions. Information about strategy and intentions or objectives can flow down to enable better control of performance and feedback about performance of intentions and objectives can flow up along with learning that helps an organization understand its capabilities and performance and learn from its experiences. Organizations that would like to roll-up or summarize performance data from business units or projects will find that quality's six Themes of Performance enable them to create assessments or measurements that are easily "normalized" – or that are expressed in common quantitative units or qualitative scales that can be more easily communicated, compared and used for monitoring and controlling strategic performance.

In the Process Enrichment® philosophy, quality's six Themes of Performance serve to provide a specific way to go about understanding how to define quality for anything that you're interested in defining quality for – be it an organization or the products and services the organization produces. This is one of the reasons why we've been following the example of Mr. Churchill's We Shall Fight on the Beaches speech – it provides an opportunity to demonstrate how inseparable strategy and quality are and how strategy and quality can be defined for something as complex as the war strategy of a nation. As we progress in our exploration of the concepts of strategy, quality and risk we will attain an understanding of how these elements are inseparably related and how to use this relationship to our advantage to master the talents of strategic communication and continual quality improvement.

In the cases presented for this chapter, in the following section, we will explore strategic issues with a depth of clarity that is uniquely attained through our understanding of quality's six Themes of Performance.

Quality's Six Themes of Performance

▶ **The U.S.S. Indianapolis**
Presenting the challenge of quality's Acceptance Theme of Performance:
How will we create performance quality that's acceptable to all stakeholders?

▶ **Aluminized Explosives ("Most Secret")**
Presenting the challenge of quality's Efficacy Theme of Performance
How will we create measurably effective performance quality?

▶ **The LZ 129 Hindenburg**
Presenting the challenge of quality's Sustainment Theme of Performance
How will we sustain our desired level of performance quality?

▶ **The Battle of the Atlantic**
Presenting the challenge of quality's Optimization Theme of Performance
How will we improve and perfect the ability to perform a successful strategy?

▶ **The Final Voyage of the R.M.S. Titanic**
Presenting the challenge of quality's Process Enrichment Theme of Performance
How will the people involved in the system interact and perform best?

▶ **The Battle of Britain**
Presenting the challenge of quality's World Class Theme of Performance
How will we achieve performance quality excellence everywhere we perform?

Six prototypical case studies are presented here to provide a deeper understanding of the character and real-world significance of each of quality's six Theme of Performance. The cases selected are significant historical episodes that exemplify the type of consequences that, on a more or less significant scale, can occur when an organization, a project, a process, a product, a service, or any other "Performance Unit," bears a significant risk of poor performance in one or more of the Themes of Performance. The term "Performance Unit" is used in some of the case studies as a generic way to refer to any thing or idea whose quality or ability to perform in accordance with strategy is being described.

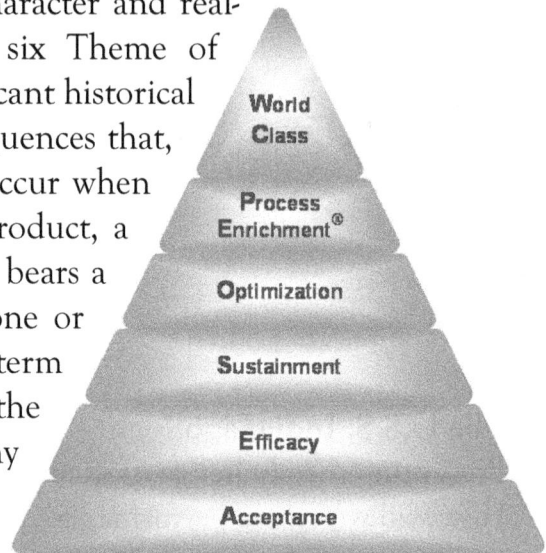

Properties of Quality's Six Themes of Performance
_____

1) Each of the six Themes of Performance represents a unique theme or aspect of quality.

2) The six Themes of Performance collectively provide a comprehensive definition of quality for any Performance Unit.

3) Poor quality in a lower level Theme of Performance may increase the risk of poor performance in a higher level Theme of Performance by reducing its scope of performance. For example, a pencil without an eraser may be marginally acceptable, and it may be effective in writing, but as the user needs to erase misspellings, or other mistakes, the pencils' limitations create poor or unacceptable performance in practical usage scenarios – a concern of the Process Enrichment Theme of Performance.

4) It's not unusual for a Performance Unit to be good in some Themes of Performance and poor in others. For example, a cheap pen may be acceptable in cost, effective in writing, and comfortable to hold, but it may be unsustainable because it promptly runs out of ink.

5) To prioritize quality improvement effort, select features to improve in the Theme of Performance where higher quality is most likely to accomplish intentions or improve attainment of the desired strategic incentives – for

example, improved market share, competitive advantage, profit or safety. Empirical evidence suggests that people, generally, place greater value on prioritizing improvement effort in the Acceptance Theme first and then work upwards through to the World Class Theme continually nudging performance in the direction of quality excellence. Considerations in prioritizing improvement are a topic in chapter 4.

6) As innovation occurs or as improvements are made in high level Themes of Performance, the expectations for quality excellence in lower level Themes may evolve towards higher quality expectations. This reflects an inherent property of the quest for ascendant quality – that as improvements demonstrate an improved quality of life, the concept of quality evolves transforming desires towards higher levels of performance in an inexorable progression.

7) As a consequence of the generic nature of the definition of quality, and of quality's six Themes of Performance, these definitions create a simple, common, and deeply expressive language of quality which can enable you to describe, define, or assess the quality of any thing or of any idea with great specificity. As we will soon see, the Themes of Performance can also be used to develop targeted performance measurements for just about any thing or any idea through a relatively simple, systematic, approach. Quality's six Themes of Performance are also the basis for Saile charts, introduced in chapter 3, which can be used to simultaneously visualize strategy, quality, and risk for any Performance Unit.

Case Study:

The U.S.S. Indianapolis

The Challenge of the Acceptance Theme of Performance

- How will we create performance quality that's acceptable to all stakeholders?

Introduction

The Acceptance Theme of Performance is the first echelon in the hierarchy of quality's six Themes of Performance. The Acceptance Theme of Performance considers what it takes for an organization, product, service or any other "Performance Unit" to be acceptable to the customers or to other stakeholders for whom the Performance Unit is intended to provide benefit. The Acceptance Theme is especially significant because it is the first screen that stakeholders use to determine whether a Performance Unit has, at the least, the minimum abilities to perform that make it acceptable. If a Performance Unit is too expensive, can't meet mandatory requirements, or cannot be produced when needed, customers may find the Performance Unit unacceptable and unworthy of further consideration. A supplier or a customers' consideration of the Acceptance Theme of Performance is intended to prevent unintentional capability lapses or the omission of something required or expected by the intended end-users or customers, and other stakeholders, which would render the products or services unacceptable.

Prolog

The tragic story of the aftermath of the sinking of the U.S.S. Indianapolis, during WWII, is exemplary of the significance of the Acceptance Theme of Performance to an organization. One of the "Focus Areas" within the Acceptance Theme of Performance is called "Content Coverage." When the Performance Unit is an organization, Content Coverage is an analysis, usually supported by measurements, intended to make sure that all necessary organizational infrastructure – such as policies, procedures, templates, tools and training – is created, peer reviewed, approved, and available for use where needed. Management of Content Coverage is essential to the survival and success of any organization. For this reason, organizations are more likely to be influenced to address risks of poor performance in the Acceptance Theme of Performance than in the higher echelon Themes of Performance. It was a lapse in management of Content Coverage that turned the disastrous sinking of the U.S.S. Indianapolis into a tragedy.

The U.S.S. Indianapolis

The U.S.S. Indianapolis was a U.S. Navy heavy cruiser serving through the Second World War. The U.S.S. Indianapolis was active in support of convoys and invasion forces and is also noted for transporting the materials for the first atomic bomb used in warfare to its staging area. The U.S.S. Indianapolis' service ended tragically, just two weeks before the end of WWII, in a manner described on a Web site developed by two grandchildren of one of the U.S.S. Indianapolis' survivors.

"At 14 minutes past midnight, on 30 July 1945, midway between Guam and Leyte Gulf, she [the U.S.S. Indianapolis] was hit by two torpedoes fired by the I-58, a Japanese submarine. The first blew away the bow, the second struck near midship on the starboard side adjacent to a fuel tank and a powder magazine. The resulting explosion split the ship to the keel, knocking out all electric power. Within minutes she went down rapidly by the bow, rolling to starboard.

Of the 1,196 aboard, about 900 made it into the water in the twelve minutes before she sank. Few life rafts were released. Most survivors wore the standard kapok life jacket. Shark attacks began with sunrise of the first day and continued until the men were physically removed from the water, almost five days later."[50]

ussindianapolis.org

It is the unnecessary five day delay in the rescue of the 900 sailors of the U.S.S. Indianapolis, which compounded the loss of 296 lives, due to the ship's sinking, to a loss of 879 lives as another 583 sailors perished while awaiting rescue. During those five days, stranded in the ocean, sailors succumbed to

---

[50] http://www.ussindianapolis.org/story.htm, retrieved on 3/14/2012.

injuries, lack of food and water, drowning, shark attacks and, eventually, wild attacks by fellow sailors who developed terrible hallucinations after drinking seawater.

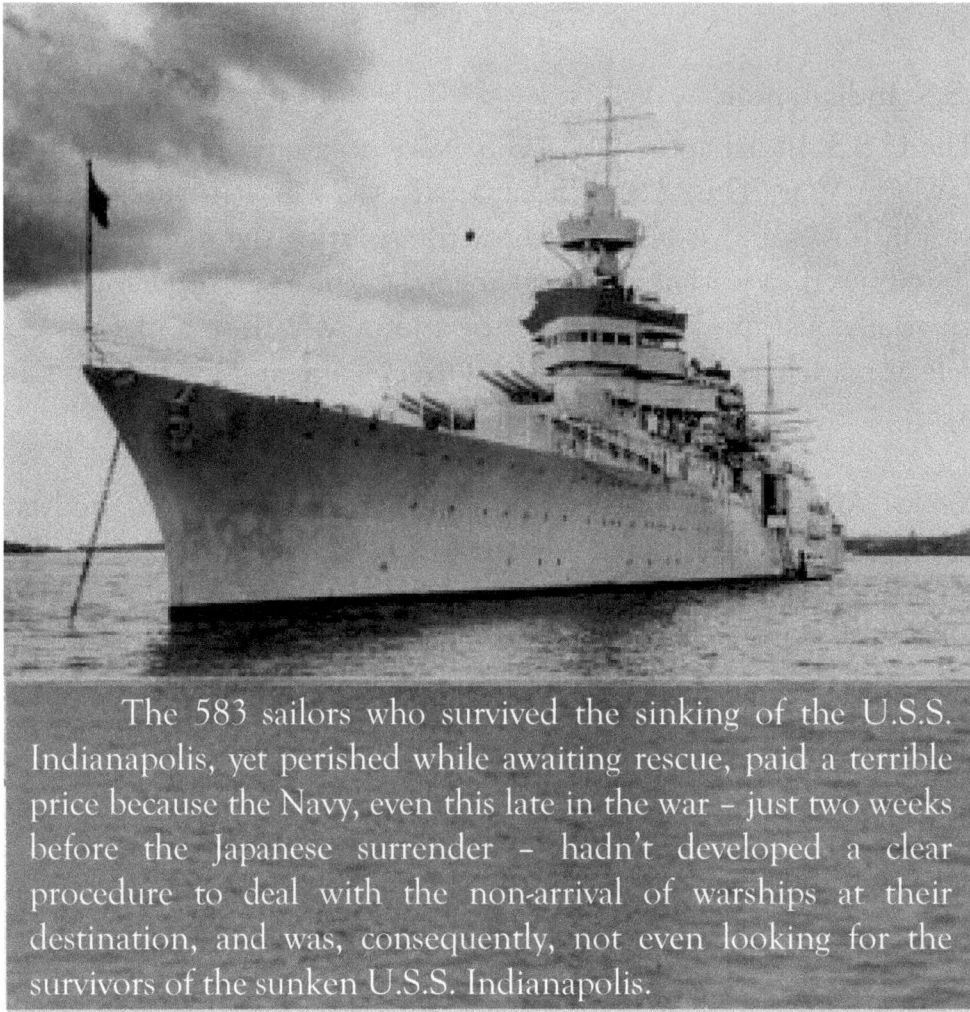

The 583 sailors who survived the sinking of the U.S.S. Indianapolis, yet perished while awaiting rescue, paid a terrible price because the Navy, even this late in the war – just two weeks before the Japanese surrender – hadn't developed a clear procedure to deal with the non-arrival of warships at their destination, and was, consequently, not even looking for the survivors of the sunken U.S.S. Indianapolis.

Figure 2-6, The U.S.S. Indianapolis, and one significant cause of the tragedy befalling the survivors of the ship's sinking.

The Navy's explanation for the five day delay in responding to the sinking of the U.S.S. Indianapolis is that in January, 1945, the Navy, "In order to reduce the volume of radio traffic and increase security," issued a directive that "made specific provision that the arrival of combatant vessels was not to be reported." Although "nonarrivals were expected to be reported" the Navy states that an incorrect inference was drawn that "since arrival reports were not to be made for combatant ships, by the same token neither were reports of non-arrivals to be made." The Navy spoke of this incorrect inference, or, rather, this absence of a clear procedure, as "a weakness in the organization." Indeed, this failure is an organizational failure because it was incumbent on the Navy to have an organizational procedure to recognize the need to rescue the crews of sunken warships during wartime – especially since this was a very foreseeable, very critical, stakeholder need.

There were, reportedly, other lost opportunities to recognize that the U.S.S. Indianapolis was overdue at its destination, however, this particular cause of failure is of interest because it is prototypical of the form of organizational failure that may result from a risk of poor performance in the Acceptance Theme of Performance. Failures that happen in the absence of documented procedures may appear to be caused by an absence of common sense, or reliance on poor assumptions, however, it should be appreciated that establishing common sense and preventing poor assumptions in an organizational setting is the purpose of the body of documented organizational procedures. When common sense, especially for predictable, often repeated, and critical activities hasn't been integrated into procedures, this circumstance can contribute to poor judgment.

The 317 survivors of the U.S.S. Indianapolis tragedy owe their lives to the crew of a PV-1 Ventura Bomber on routine antisubmarine patrol who, by chance, noticed the oil slick from the sunken U.S.S. Indianapolis and circled in for a closer look – leading to their rescue.

Epilog

While the story of the U.S.S Indianapolis is outstanding in the tragic loss of life, if we accept that the cause of the extent of loss of life was due to the insufficiency or omission of procedures for a predictable stakeholder need – the need to follow-up on missing warships and to perform search and rescue operations, then we are presented a strong lesson, which we may recognize as a potential risk of poor performance affecting any organizations' performance or any performance Unit's performance – when essential stakeholder needs are not met. The proper assessment of Content Coverage is key to avoiding the serious form of performance lapses that are characteristic of failures in the Acceptance Theme of Performance.

The important Content Coverage questions to consider are "What doesn't this Performance Unit do that it should be capable of doing?" and "Which customers and other stakeholders aren't being adequately served by the Performance Unit?" When deficiencies are recognized, transformation plans should be made to track progress in addressing the deficiencies until all is well. Simple and powerful Content Coverage measurements can address the risk of poor performance in the Acceptance Theme of Performance.

Organizational Implications

In an organizational setting there are many ways to track Content Coverage. A spreadsheet can be used to list the names of procedures in varying stages of development and the status of their acceptance for use. A peer review process is very effective in enabling stakeholders to view and comment on a procedure until necessary and appropriate content has been drafted, finalized, and approved for use. Desired, but not yet drafted, procedures should be tracked by the same system in order to increase their visibility and to take advantage of resource availability to perform the creative effort of producing the desired procedures when possible. Popular or frequently used business rules are the most likely candidates to become new procedures. In organizations that follow standardized sets of processes like the CMMI or ISO 9000 series it is all too common to leave out procedures that are necessary but that are not parts of the "model" and that aren't on anyone's checklist. For example, how to prepare a time card; how to request a computer for a new employee; who to notify when you're out sick; etc.

Case Study:

Aluminized Explosives ("Most Secret")

The Challenge of the Efficacy Theme of Performance

- How will we create measurably effective performance quality?

Introduction

The Efficacy Theme of Performance is the second echelon in the hierarchy of quality's six Themes of Performance. Efficacy in an organization, product, service, or in any other "Performance Unit" is valued because it is closely associated with the ability of a Performance Unit to attain the desired incentive of a customer's strategy. For example, if the incentive of our strategy is to attain a cure for a disease, we may view the efficacy of a medicine to be the decisive form of quality guiding our selection of a medicine. The efficacy of a medicine in curing a disease can be measured by whether or not the disease is cured in patients taking the medicine. If there are alternative medicines which are measurably effective in curing the disease, the speed at which each medicine cures the disease is likely to be an additional aspect of efficacy that would serve as a basis for competition. Efficacy provides a strong competitive differentiator signifying the potential for success a medicine offers as part of a strategy to cure a disease. Efficacy is often prominent as a basis for comparison, not only in medicines but in other products and services as well, which makes awareness of the efficacy of competing products and services a key consideration in design and development or in acquisition of creative effort.

Prologue

Efficacy in the service of customer strategy often stands out as the form of quality with decisive value in many competitive endeavors. For this reason, being content, for too long, with good efficacy, when quality excellence in efficacy is necessary, can create a significant risk of poor performance. Even quality excellence in efficacy is an ideal which is driven to evolve over time by causes such as improved competing technology, changes in the way a product or service is used, and changes in aspirations that raise customer's expectations. When an organization has the ability to understand customer strategy and how their customers ideal of quality excellence evolves over time, they may use this insight, and their talent in innovation, to transform their products and services to a form which can secure continued quality excellence or to a form which can serve to establish a new ideal of quality excellence for their targeted customer market. Once a product or service gets beyond competing as a commodity, which is only expected to conform to basic requirements, efficacy becomes prominent as the differentiating form of quality and it becomes increasingly important for the forms and degrees of efficacy to be managed in order to match the abilities to perform which are in demand and are necessary to serve a customer's strategy.

A lesson in the significance of managing the Efficacy Theme of Performance, in order to attain strategic incentives, resonates from an episode dating from the midst of the Second World War.

Aluminized Explosives

During the Second World War, Frederick Lindemann, who was a scientific advisor to Britain's Prime Minister, Winston Churchill, attended a briefing, where several munitions experts discussed the relative power of German and British bombs. These munitions experts made forecasts of the ability of the respective nations' bombing missions to cause damage to buildings and to other infrastructure. Mr. Lindemann heard that the experts were ascribing nearly twice as much power to German bombs as to British bombs because the Germans mixed aluminum powder with the high explosive in their bombs and the British didn't. When Mr. Lindemann brought this peculiar information to Mr. Churchill's attention, Mr. Churchill ordered that a searching inquiry be made, under Mr. Lindemann's guidance, to uncover the causes of this circumstance. Mr. Churchill also ordered that "The whole matter is to be kept most secret."[51]

Mr. Churchill disclosed the facts of this secret episode to the public, a few years after the war, in a volume of his set of books on the Second World War titled "Closing the Ring." Mr. Churchill wrote that when the findings of Mr. Lindemann's inquiry were discussed "The results astonished everyone concerned."

---

[51] Churchill, Winston. "*The Second World War: Closing The Ring,*" Vol. 6, (Houghton Mifflin, 1951) p. 525.

"It appeared that in the early days when aluminum was scarce it had been decided to use all the aluminum powder which could be spared for making depth-charges, and that this custom had persisted, although aluminum had now become more plentiful. Orders were immediately given to improve our explosive - in the first place in our heavy bombs - by adding aluminum powder, and their efficiency during the whole latter half of the war was thus increased by about half as much again."[52]

Winston Churchill
1951

Epilogue

The British were, understandably, astonished by the troubling discovery that they had gone through half of the Second World War without realizing that the bombs they were using were only 2/3 as potent as they could have been. Many lives were put at risk to fly the bombers carrying the bombs over their targets through enemy antiaircraft fire and through a gauntlet of enemy fighter planes. Even more significantly, many thousands of lives could have been saved for each day that the increased "efficiency" of the bombs and, therefore, the efficacy of the bombing missions could have shortened the war. This episode was particularly disquieting because it revealed that a terrible, yet avoidable, failure in the management of efficacy had occurred – a fact which, if the public was aware of it, could have been embarrassing and damaging to Britain's military leadership. It was an episode which was, perhaps, well worth being kept "most secret."

In a proper scheme of quality management, the risk of poor performance in using bombs without aluminum powder should have been tracked as a risk of poor performance[53], in the Efficacy Theme of Performance, until it was possible to restore the desired level of quality – when aluminum powder became available.

---

[52] Churchill, Winston. "*The Second World War: Closing The Ring,*" p. 525.

[53] Readers familiar with the last century's risk management practices may have been influenced to believe that once a risk is "realized" it becomes an "issue." This is a counterproductive notion since these forms of "issues" often maintain their risky nature and are, in fact, risks. Masking a risk as an issue may free the risk of the additional creative reactions such as design changes, implementation enhancements or mitigation steps and controls which serve to prevent undesirable consequences from occurring.

Whether performance deficiencies are tracked as deviations from quality excellence or as risks of poor performance, publicizing these deficiencies can influence corrective action and improvement that's necessary to remain competitive and that's necessary to avoid the possibility of undesirable or even tragic consequences. The capability of organizational management to have visibility into the relationship between current performance, competing performance, and the standard of quality excellence is a capability that can serve to raise the quality of an organization to the level of quality excellence and beyond[54]. Mr. Churchill's use of Mr. Lindemann and a variety of experts to obtain feedback on performance, in many areas where performance was critical to success, is the broader lesson of the Aluminized Explosives story which Mr. Churchill conveyed in his book.

> "This episode shows how useful it is in great organizations to have a roving eye."[55]

> Winston Churchill

By proper evaluation and management of performance, in each of quality's six Themes of Performance, for all forms of creative effort, an organization can position itself for its best prospect of strategic success and for its best contribution to the progress of civilization.

Organizational Implications

The challenge of creating efficacy, as presented in this case study, required only that the players in our story understand the need to introduce aluminum powder into their explosive formulation. Our case study assumed that the appropriate formulation was already known to the munitions experts – which may have been the actual case. The classic mathematical model for identifying the blend of various ingredients in a product, which optimizes the product's efficacy, is called the "feed mix problem" in mathematical programming. The feed mix problem is a best practice for any industry where products depend on a blend of ingredients. For example, optimizing the types of ingredients of various

---

[54] In chapter 3 we will discuss a methodology called Saile Charting which offers a capability to visually communicate the relative performance of competing products and services against a standard of quality excellence in order to manage and control strategic performance.

[55] Churchill, Winston. *"The Second World War: Closing The Ring,"* p. 526.

nutritional values and costs in chicken feed as a feed mix problem can make a big difference in the health, weight, and cost of the chickens raised. As another example, the blend of chemicals used in gasoline can be optimized for power and for cost and to minimize undesirable emissions, etc. The feed mix model is designed to simultaneously optimize a set of resource constraints in order to maximize the value or efficacy of a formulation – based on an "objective function." Used creatively, the model can also be used to optimize a blend of products and services as well. There are several good examples of feed mix problems available on the internet. Solving Feed Mix problems is made easier by readily available software such as Microsoft Excel.

To improve the efficacy of almost any organization, we can improve the documentation such as policies, procedures, templates, tools and training which the organization uses to engage its various competencies. Since these organizational documents and resources are intended to bring consistency and professionalism to the performance of the tasks they support, the less these resources need to be tailored – or worked around – by the people using them, the better their efficacy is. When tailoring is found it should be considered an opportunity to improve. The improvement effort may entail investigating performance to see why the tailoring was necessary. If the rationale for the tailoring was sensible, or if the tailoring was the best way to perform, then the affected documentation or tools should be updated to integrate the better way of performing. On the other hand, when undesirable deviation from the desired best practice is found this may indicate a need for exposure to strategic communication that educates people on the benefit of the established practices. Tailoring can be easily measured – though it may seem odd to measure something with a goal of zero tailoring. Efficacy can also be loosely indicated by the frequency of voluntary use of organizational resources. This frequency may be easy to obtain if the resources are maintained on a Web page. Missing policies, procedures, templates, tools and training may first be noticed as an efficacy problem, however, once the need for a resource is identified this becomes an Acceptance Theme risk of poor performance. When the effect on efficacy of a missing resource is minor, people will tend to create business rules to work around the risk, however, if the risk is frequently encountered or impactful this should influence transformation of the business rule into a more formal policy or procedure.

Case Study:

The LZ 129 Hindenburg

The Challenge of the Sustainment Theme of Performance

- How will we sustain our intended degree of performance quality?

Introduction

The Sustainment Theme of Performance is the third echelon of quality's six Themes of Performance. Consideration of the Sustainment Theme is intended to ensure that a Performance Unit has an operational capability to perform which is sustained over the Performance Unit's desired useful life. A supplier's development team or a customer is likely to focus on sustainment once an acceptable and efficacious Performance Unit has been successfully envisioned. Sustainable performance is continually valuable, reliable, and maintainable, and efficiently uses a customer's effort in support and management. When a product or service is unsustainable it can lose its valuable ability to perform or it can fail unexpectedly at an inopportune time. This is especially significant when the failure can result in injury or in loss of life to customers. Unsustainable performance is also a source of early defects or complaints which are likely to cause customers to reevaluate their strategies so they can use alternative products and services to attain better performance. Poor cost management and the wasteful use of resources are also risks of poor performance in the Sustainment Theme because organizations must be financially, as well as environmentally responsible in order to secure organizational continuity.

Prologue

When any of quality's six Themes of Performance are inadequately considered in the design of creative effort, and the creative effort doesn't achieve quality excellence, this risk of poor performance travels as an inherent part of the product or service wherever it goes. Slight risks are usually acceptable and may even be desirable since, as we discussed with regard to planned obsolescence, under the right circumstances, acceptance of some degree of risk can reduce labor and material costs and can be highly profitable. On the other hand, risk taken in the sustainment of the ability to perform safely should be considered especially dangerous because this is risk that bears an ever-present potential for disastrous consequences. In the case of safety risks, concern for the moral and ethical impact must be appropriately balanced in the decision to produce or use a product and service.

The LZ 129 Hindenburg

Figure 2-7, May 6, 1937, the LZ 129 Hindenburg erupts in flames at the Lakehurst Naval Air Station in New Jersey.

The LZ 129 Hindenburg airship was named after former German president Paul von Hindenburg. The Hindenburg was designed to transport passengers in luxurious conditions during local and overseas voyages. The Hindenburg airship was kept aloft by a massive amount of hydrogen gas contained within its airframe. As shown in Figure 2-7, on May 6, 1937, at the end of its transatlantic journey, the Hindenburg caught fire as it was docking at its landing site and burst into a deadly fireball. The Hindenburg disaster was captured by news cameras as it happened. The event was famously dramatized by Herbert Morrison, a shocked radio announcer, who was delivering a live broadcast of the arrival of the Hindenburg as it maneuvered towards its mooring.

There has been a great deal of speculation about what caused the hydrogen gas in the Hindenburg to leak and there has also been speculation about what caused the hydrogen gas to ignite. In any case, we know that an explosion did occur unexpectedly and at an inopportune time. These are characteristic signals that the failure was the outcome of the risk of poor performance in the Sustainment Theme of Performance. Regardless of the many ways the actual ignition of the hydrogen gas could have occurred, the failure was caused by the strategy of using flammable hydrogen gas for buoyancy instead of using nonflammable helium gas. The Hindenburg's passengers were probably ready, willing, and able to pay for use of the more expensive helium gas; however, in the 1930's, helium was difficult to obtain in the volume needed. The U.S. was the leading supplier of helium, at the time, and had a ban on export of the gas to Germany – due to deteriorating political conditions as Germany descended into National Socialism. Considering their circumstances, and their ethical dilemma, the manufacturers of the Hindenburg decided to design the Hindenburg to run on the cheaper, more readily available, and more buoyant – but dangerous – hydrogen gas accepting the risk of poor performance in the Sustainment Theme of Performance.

In hindsight, from a contemporary perspective, it may have been better for the builders of the Hindenburg to relocate their facilities to a country which could provide helium as well as the talented craftspeople to build the airframe. This would undoubtedly be a costly prospect that may have been very impractical or impossible at the time. Faced with the ethical dilemma of either shutting down their business until political conditions got better or accepting the risk of using dangerous hydrogen gas, the builders ran the risk and their customers and employees paid for that decision with their lives. Ethical dilemmas are often difficult to resolve, however, when the Sustainment Theme is properly considered, it becomes more likely that the necessity of resolving them is clear – which can influence and enable people to appropriately identify and consider alternatives.

Epilogue

Although the Hindenburg integrated new technology, was luxurious, fast, and could be considered acceptable and efficacious to a high degree – it wasn't sustainable because it was mostly made of hydrogen – a combustible gas – and its regular use put it in circumstances where the gas could ignite. The use of Hydrogen for buoyancy, instead of helium, was a risk of poor performance in the Sustainment Theme of Performance which the Hindenburg carried with it wherever it went. Ultimately, this risk of poor performance in the Sustainment Theme brought about the Hindenburg's downfall. The Hindenburg's form of failure – which was unexpected and at an inopportune time – is characteristic of failures in the Sustainment Theme.

Organizational Implications

Failures in the Sustainment Theme of Performance can occur gradually and can remain masked unless someone is looking for them. For example, a mail order company that hasn't properly marked up the price of their products to account for shipping costs may be siphoning off profits from their products. For this reason an accurate accounting should be made of expenses and revenue. If the undercharging would tend to drive the company out of business, this would be characteristic of a risk of poor performance in the Sustainment Theme.

For products or services which tend to fail during use, at the end of their useful life, it is desirable to have a signaling mechanism to warn the user of the impending failure. For example, a routine brake inspection lets drivers know, well ahead of time, when they need to replace their car's brake pads. For some products, measures of reliability can be used to construct a maintenance schedule or to signal when to replace expended resources in order to sustain the desired operational capability. Understanding how a product fails may also require destructive testing of a product sample. This may be the best way to discern the knowledge of product characteristics necessary to determine the best course of action to prevent an undesirable lapse in a products ability to perform successfully.

Case Study:

The Battle of the Atlantic

The Challenge of the Optimization Theme of Performance
_____

- How will we improve and perfect the ability to perform a successful strategy?

Introduction

The Optimization Theme of Performance is the fourth echelon in the hierarchy of quality's six Themes of Performance. Consideration of the Optimization Theme of Performance reveals how well an organization, product, service, or any other "Performance Unit," has matured and perfected the design-best implementation of its strategy. Optimization of the highly valued aspects of performance which are already acceptable, efficacious, and sustainable may be of significant influence to customers as a competitive discriminator. Optimization can relate to any aspect of the ability of a Performance Unit to perform a customer's intended strategy and, for this reason, implies refinement in the technical performance in any one or more of the six Themes of Performance. Expressions such as "the precision of a fine watch," "running like a fine tuned instrument," "Honed-in on a great idea," "personalized service," etc. reveal how commonly the ability to perform optimally is appreciated and valued for a wide variety of creative effort.

Prologue

The story of the long and lethal battle raging throughout World War II, which Winston Churchill referred to as the Battle of the Atlantic, is a true story of a competition between adversarial strategies. On the German side of this battle, the German navy sought to sustain U-boats (submarines) in active combat against all forms of British shipping. On the British side, the British navy and air force sought to protect their vital shipping and sea linkage to their allies by sinking the U-boats. In these competing strategies, the essential incentives of sinking each other's ships and surviving to win the war remained consistent; however, the Battle of the Atlantic was repeatedly transformed by the optimization of defensive and offensive strategies by both sides. Each adversary used innovation and refinement of technical performance to optimize and improve the survivability of their own ships and to improve the lethality of attacks on their adversary's ships. As new or newly optimized strategies emerged, they created risks which had to be circumvented or counteracted by the best strategy, using the best technology, which each adversary could devise. As the ability to perform their intended strategies, on a world-wide scale, became more or less risky, the trend towards victory in the Battle of the Atlantic shifted back and forth between the adversaries.

The Battle of the Atlantic

Figure 2-15; April 17, 1943, Coast Guardsmen on the deck of the U.S. Coast Guard Cutter Spencer watch the explosion of a depth charge which sank the German U-boat U-175. This image was provided by the U.S. National Archives and Records Admin.[56]

The longest battle of World War II was the battle at sea which Winston Churchill called the Battle of the Atlantic. Germany presented itself at the beginning of World War II with a formidable U-boat fleet. U-boats distinguish themselves from other forms of warship by their utilization of stealth – hiding in

---

[56] NARA, figure 26-G-1517, Retrieved on 6/1/2013;
http://www.archives.gov/research/military/ww2/photos/images/ww2-61.jpg.

a low profile when operating on the surface and hidden undetectably to the human eye when operating underwater. All British shipping became a target of U-boat attacks – which were initially extremely difficult to defend against because the U-boats' stealth facilitated their attacks and their escape after their attacks. Since the British island imports much of its food, attacks upon British shipping were particularly troubling. Britain also depended on shipping for supplies of armaments necessary for war. Germany needed to destroy British shipping to reduce the capability of the British to stay in the war and also to reduce the capability of the British to bring troops and supplies to their allies on the continent to attack Germany. As losses in shipping mounted for the British, they developed a convoy system to offer some protection to their ships. In response, the German navy developed Wolf Packs to attack convoys, systematically, in a way that was orchestrated to overwhelm defensive resources. As the British became better at locating and destroying U-boats, using new and evolving radar technology, the time that the U-boats could remain exposed out on the ocean surface, and active in their attacks on shipping, shortened. The development and improvement of SOund Navigation And Ranging (SONAR), which is used to locate submerged U-boats also increased the risk to U-boats and, like the improvements in radar, counteracted the effectiveness of their stealth. To extend the operating time at sea the German navy built tanker U-boats which refueled the other U-boats. Radio communication providing missions and other information to Germany's U-boats at sea was encrypted by a device called an Enigma machine. In 1939, the British were given, by their Polish allies, methods to decode messages encrypted with the Enigma machine which were discovered by three polish mathematicians, Marian Rejewski, Jerzy Rozycki and Henryk Zygalski of the Polish Cipher Bureau. The Polish Cipher Bureau also provided Polish-made replicas of Enigma machines to the British.[57] As Polish mathematicians worked with the British code breakers at Bletchley Park to decipher the Enigma code they became able to determine the location of the tanker U-boats so they could be specifically targeted. The British were also influenced by a talented mathematician of their own who initiated a form of applied mathematics which came to be known as Operations Research. Through Operations Research, the British determined that, in aerial attacks, their "best chance of killing a U-boat was to concentrate on the ones that were surprised at

---

[57] Wilcox Jennifer, *Solving the Enigma: History of the Cryptanalytic Bombe*, Center for Cryptologic History, National Security Agency, Revised 2006, p. 8.

the last few seconds and had dived only a few feet, leaving them without time to change course."[58] Operations researchers also determined that the high contrast of the black paint on the bottom of aircraft against the sky allowed U-boats to see them coming from great distances. This enabled the U-boats to dive before the aircraft arrived overhead to drop depth charges. The aircraft bottoms were then painted white reducing the risk of poor performance and improving attack results dramatically. Operations Researchers also convinced the British to change the depth at which their Depth Charges were set to explode from 75 feet deep to only 25 feet deep – their minimum setting. These were effective changes after which "the lethal effectiveness of air attacks on U-boats increased by a factor of five."[59] German efforts to improve the lethality of their submarine attacks came in the form of several improvements in the design of torpedoes. An effective improvement was in the use of acoustic torpedoes which were able to travel straight and then change direction towards the noise made by targeted ships. As this improved strategy became known, "the Allies countered acoustic torpedoes with noisemaker decoys such as Foxer, FXR, CAT and Fanfare."[60] As the British increased their understanding of the risks of poor performance in their adversaries' strategies, as well as in their own strategies, they became able to optimize their own strategies and they became able to increase the risk of poor performance in their adversary's strategies and thereby win the Battle of the Atlantic.

Epilogue

The once glorious role of submariner, which had inspired sailors in the German Navy, became a dreaded role as its devastatingly high mortality rate became evident. In World War II Magazine, Mr. Stephen Budiansky wrote of the effect of operations research on the results of the Battle of the Atlantic.

[British operations researchers,] "...applying little more than a scientific mindset and a few simple equations, were able to propose

---

[58] Budiansky Stephen, *Dead in the Water: The Battle of the Atlantic turned when Allied scientists joined the hunt for U-boats*, World War II Magazine, September/October 2010. Retrieved on 6/7/2014 from: www.ww2hc.org/articles/or_atlantic.pdf

[59] Budiansky Stephen, *Dead in the Water*

[60] Bluebird Marine Systems Ltd., *U Boat Armaments*, retrieved on 6/7/2014 from: http://www.bluebird-electric.net/submarines/u_boats_german_submarines_world_wars_one_and_two.htm

seemingly tiny changes in tactics or strategy that doubled or tripled the lethality of the Allied antisubmarine campaign in a single leap."[61]

Stephen Budiansky
World War II Magazine, 2004

Process and product improvement and optimization is often mistakenly believed to be exclusively dependent on statistical measurement techniques. In the Battle of the Atlantic, we can see that very effective improvements in performance were, in several important instances, based on changes to strategy that were driven by a scientific mindset which recognized risks of poor performance in enemy strategy. For example, although the change in depth at which the depth charges were set to explode benefitted greatly from strategic considerations and statistical thinking, it was a radical change from a 75 ft. depth of detonation to a 25 foot depth of detonation which made the dramatic impact. The depth of explosion was subject to further modest improvements through statistical analysis – however – the recognition of the need to dramatically alter the depth setting was based only on analytical optimization of the offensive strategy.

The simplicity of the mathematics, in some improvements, stands out because it proves that anyone who pursues strategic improvement need not be a statistician. To change the depth setting of the depth charges to their minimum setting required only the strategic insight that success was more likely when a U-boat was caught near the surface. This solution didn't require the steady state conditions or historical record necessary for statistical analysis. The solution became self-evident once the Optimization Theme of Performance was adequately considered in the context of the enemy's strategy for escaping a depth charge attack.

---

[61] Budiansky Stephen, *Dead in the Water*

Organizational Implications

Optimization can be in great demand when customers are capable of appreciating the benefit of a mature and perfected strategy for performance of their desired strategies. Optimization can be achieved very effectively by careful analysis of strategy. A starting point for optimization is the careful, creative evaluation of each element of strategy and each Theme of Performance. Usually, the capabilities and resources that are readily available or easily acquirable can provide a focused set of improvement opportunities which will be relatively easy to contemplate in new and beneficial ways. Following this exploration, conceptual ideas can be developed identifying completely new ways to attain the incentives desired. Once an optimal strategy is identified, further optimization of the implementation should be considered. Mathematical Programming techniques are best practices in optimization. Although techniques of Mathematical Programming have been in use since their invention in Russia during the Second World War, they are still undiscovered by most organizations – even though they have never been easier to use than they are today with common computer applications like Microsoft Excel.

In an organizational setting, assuming infrastructural documents like policies, procedures, templates, tools, and training are acceptable, efficacious, and sustainable, they can be optimized by measuring and controlling their reading Ease, and grade level – to make them easy for employees to understand.

Case Study:

The Final Voyage of the R.M.S. Titanic

The Challenge of the Process Enrichment Theme of Performance

- How will the people involved in the system interact and perform best?

Introduction

Quality in the Process Enrichment® Theme of Performance is an indication of how well an organization and an organization's products and services are dialed-in to the one best way of performing for the people in the system. When customers implement their strategies in real-world circumstances their desire is for the products and services they've acquired to perform successfully in every relevant respect. Customers also expect their supplier organizations to exhibit professionalism in their services and to be dependable partners when questions or issues arise. For suppliers, achieving quality excellence in the Process Enrichment Theme of Performance involves meeting three very prominent and interrelated challenges:

- A technological challenge in identifying product and service features that are consistent with performance quality excellence in the targeted customer groups' real-world usage scenarios;
- A human factors challenge in making the behaviors and interactions that are necessary for a customer's operational strategy comfortable, safe, effective and practical – to the extent that customers are motivated to acquire and use the strategic solution offered; and
- An organizational planning challenge in creating or acquiring the capabilities necessary to produce competitive, technologically well matched, and desirable products and services.

These challenges, in turn, require that the employees of the supplier organization attain appropriate rewards for their investment of creative effort into the strategic solution. For a strategic solution to attain its best opportunity to be crafted in the one best way of performing there must be common ground, common benefit, and, therefore, harmony between the supplier's, the customer's, and the creative artisan's independent strategies. Attaining this Coascendancy can be an inspiring as well as a demanding creative challenge calling for the best of collaboration among all the people in the system. When people in a system act in mutually beneficial ways, and in ways in which they enjoy interacting, Process Enrichment has succeeded.

Prologue

The goal of creating products and services that customers are motivated to demand challenges a supplier to demonstrate expertise in their craft in order to present customers a trouble-free and enjoyable usage experience under real world conditions. This experience of trouble-free performance must arise from creative effort which robustly supports the customers' ideal usage scenarios through quality excellence in each Theme of Performance. In products and services that achieve quality excellence, technology is not merely for show – customers expect automation and technology to be integrated, to the extent possible, to help them attain the incentives of their strategies more easily, more reliably, and faster – when possible and when safe to do so. When the technology of a product or service is not relevant to a customer's intended strategy, in the manner in which the customer expects it to be, performance is likely to be unacceptable, inefficacious, unsustainable, or suboptimal in some significant way, and the peaceful enjoyment of the creative effort and the ideal of quality excellence cannot be achieved. Under these circumstances, the risk of poor performance may lurk just beneath the surface and the pride in technological achievements, which are irrelevant to the customer's strategy, may be in the form of the arrogant pride, or hubris, of which tragedies are made.

Figure 2–8; Titanic moored in Southampton, England, April 1912.

One could imagine the pride J. Bruce Ismay, Managing Director of the White Star Line, must have experienced as his organization's creative effort, the largest passenger ship the world had ever seen, left shipbuilding firm Harland and Wolff's dock in Belfast, Ireland to sail the world's seas.

The RMS Titanic was a magnificent ship by the standard of the most luxurious ships at any time in world history. She, and her sister ship, the Olympic, embodied the shipbuilding industry's crowning achievement in first class luxury. Mr. Ismay said of the agreement with Harland and Wolff to build Titanic "We build no ships by contract at all... All we ask them to do is to produce us the very finest ship they possibly can." In Britain's post-Edwardian era, marked by the growth of industrialization and a trend towards higher social mobility, and in the United States' Progressive Era, similarly marked by social reforms and modernization and by opportunity for immigrants, the times were right for inexpensive as well as for luxury travel on the transatlantic route. This was well proven by the success of the White Star Line's competitors at the

Cunard Line, who produced the RMS Mauritania, in 1906, at 790 ft. long, and the Lusitania, in 1907, at 787 ft. long. These fine ships, although they were slightly faster, were quite outdone in terms of luxury and size by Titanic, which was 883 feet long. In this bold new age of sea travel, it was known that a fine ship could generate fine returns.

On Titanic's maiden voyage, she was well appointed with a crew of 885, to attend to 325 first class, 285 second class, and 706 third class passengers – 1,316 passengers in all. One can assume that, for one-way tickets costing as much as $100,000, in today's dollars, first class passengers benefitted most from the crew's services. The steerage, or third class fare, was only a few hundred of today's dollars – with the second class fare about double that – making the trip accessible in cost to immigrants and business travelers. Although they were not allowed to mix with passengers travelling first class, all travelers were roughly equal in bearing the risks of travel on the high seas. It was in the manner Titanic was perceived by customers to have addressed these critical performance risks that Titanic held its key competitive advantage.

A commitment to safety gave rise to the design idea implemented in the Titanic that the vessel would be capable of remaining afloat even if four of her sixteen watertight compartments with watertight, electronically controlled, doors were flooded. On this basis, a publicity brochure, produced in 1910 for the twin ships Olympic and Titanic, stated that "as far as it is possible to do so, these two wonderful vessels are designed to be unsinkable."[62] The belief that Titanic was unsinkable was probably very effective in attracting passengers from the competition. One passenger, Thomson Beattie, who "originally planned to sail home to Canada on the Mauretania"[63] sent his mother a postcard which read "We are changing ships and coming home in a new unsinkable boat."[64]

As an additional safety measure, Thomas Andrews, the chief architect and designer of Titanic and Albert Ervine were part of Harland and Wolff's nine man

---

[62] White Star Line Brochure, 1910. Retrieved on 6/7/2014 from:
http://home.comcast.net/~georgebehe/titanic/page2.htm.
[63] Behe, George. *George Behe's Titanic Tidbits*. retrieved on 6/7/2014 from:
http://home.comcast.net/~georgebehe/titanic/page2.htm
[64] Behe, George. *George Behe's Titanic Tidbits*.

"'guarantee group' who sailed on the Titanic in order to monitor her performance and correct any malfunctions that might occur."[65] Prior to leaving Queenstown, Ireland, fully loaded with passengers and crew, for the long journey across the Atlantic, Mr. Ervine sent a letter to his mother in which he described the vessel's safety arrangements:

> "This morning we had a full dress rehearsal of an emergency. The alarm bells all rang for ten seconds, then about 50 doors, all steel, gradually slid down into their places, so that water could not escape from any one section into the next.
>
> So you see it would be impossible for the ship to sink in collision with another..."[66]

Albert Ervine
April 11, 1912

Like Mr. Ervine, Mr. Andrews also believed that he had designed and his company had built an unsinkable boat. A passenger, Mrs. Eleanor Cassebeer, recounted that Mr. Andrews told her, of the Titanic, "even though it should be cut into three pieces it would still float."[67] This strange, proud, and fearless belief in the absence of the risk of Titanic sinking was also, unfortunately, fervently held by Captain Edward Smith. A passenger, Elmer Taylor, overheard Captain Smith tell his party "the ship could be cut crosswise in three places and each piece would float."[68] Referring to a ship he sailed a few years before Titanic, Captain Smith reportedly said "I cannot conceive of any vital disaster happening to this vessel. Modern ship building has gone beyond that."[69] Mr. Ismay, said what was probably the prevalent belief about Titanic's safety "the ship was looked upon as practically unsinkable; she was looked upon as being a lifeboat in herself."[70]

---

[65] Behe, George. *George Behe's Titanic Tidbits.*

[66] Behe, George. *George Behe's Titanic Tidbits.*

[67] Behe, George. *George Behe's Titanic Tidbits.*

[68] Behe, George. *George Behe's Titanic Tidbits.*

[69] Ballard, Robert D. and Rick Archbold, Lost Liners, Titanic. Reference retrieved on 6/7/2014 from http://www.pbs.org/lostliners/titanic.html

[70] Ismay, Bruce J. Testimony, British Wreck Commissioner's Inquiry, Titanic Inquiry Project, retrieved on 6/7/2014 from: http://www.titanicinquiry.org/

The story of the sinking of the Titanic on her maiden voyage, after collision with an iceberg, which resulted in the shocking loss of more than 1,517 lives, transforms from an accident, which *couldn't* have been avoided, to a tragedy, which *could* have been avoided, when the consequences of the belief that Titanic was unsinkable are considered. The main consequence of this belief was a fatal deviation from quality excellence in the strategy of operation of Titanic that increased the probability of collision with an iceberg to a near certainty.

The single element of Titanic's new operational strategy, that caused the collision with the Iceberg, is that Titanic was travelling at an excessively high speed through the icy Atlantic waters. The beliefs of Captain Smith with regard to the speed of Titanic, at the time of the collision with the iceberg, are unknown to the historical record[71], since the Captain went down with his ship, however, insight to the Captains' beliefs can be inferred from those of Mr. Ismay, who survived Titanic's sinking. During the British Board of Trade Inquiry, which was chartered to investigate Titanic's sinking, Mr. Ismay was questioned by the Attorney-General about what incentive or "object" he believed would justify the strategy of increasing the speed of travel to full speed, near midnight on April 14, 1912, having received numerous warnings of ice fields and icebergs in their path.

"Attorney-General: What is the object of continuing at full speed through the night if you expect to meet ice? Why do you do it?

Mr. Ismay: What is the use of doing it?

Attorney-General: Yes?

Mr. Ismay: I presume that the man [the Captain] would be anxious to get through the ice region. He would not want to slow down upon the chance of a fog coming on.

Attorney-General: So that on a perfectly fine, clear night, with the expectation that you are coming within the region of ice, your view is that the Captain would be justified in increasing his speed?

---

[71] Officer Lightoller testified that Captain Smith said to him "If it was in a slight degree hazy there would be no doubt we should have to go very slowly." This only speaks to a condition in which Captain Smith would slow Titanic down. An expression of Captain Smith's beliefs about running Titanic, at full speed, in the dark of night through the icy region of the Atlantic – although this is what the evidence shows he actually did – is not available.

Mr. Ismay: I do not see any reason why he should not, so long as he could see sufficiently far to clear the ice."[72]

This strange, proud, and fearless belief in the ability of Titanic to venture full speed through the region of ice is inconsistent with the beliefs which were perhaps typical of ships' Captains at the time. An article in the June 26, 1915 edition of the New York Times summarized the testimony of Captain Robert Niss of the steamship Bohemia who testified in a legal proceeding regarding Titanic's sinking. "Even on a clear, starlight night, he said, he could not guarantee to see ice far enough ahead to avoid striking it. Even on a bright moonlit night, unless the moon was behind the ice, outlining it, ice might easily be overlooked. If he had entered an ice zone described to be similar to that under which the Titanic met her fate, he said he would have slowed down to a speed which made it certain that he could stop in an emergency."[73]

Perhaps the best explanation for Captain Smith's deviation from the strategy of cautious, reduced speed, travel which common sense and the common practices of other Captains would dictate, under the circumstances, was the arrogant pride of Hubris. Hubris was the human weakness, engendered by the belief that Titanic was unsinkable, that enabled Mr. Ismay, the Captain, and members of the crew, to mistake the technology necessary to survive an accident for the technology necessary to travel faster under the nighttime conditions of the icy Atlantic. In the perspective of quality's six Themes of Performance, watertight compartments with watertight electronically controlled doors is the technology of the Sustainment Theme of Performance because this system provided the ship the ability to perform a strategy of floating until repairs or rescue could be effectuated. In the Process Enrichment Theme of Performance, if it was Mr. Ismay's intention to improve the ability of the ship to clear the region of ice more quickly, the technology necessary to do so would involve innovations such as radar or sonar or satellite tracking, which could give advanced warning of icebergs, and strong steel alloys especially configured in Titanic's hull for icebreaking. These technologies may have improved the ability of Titanic to travel at a higher speed; however, they had not been invented in 1912. In the belief

---

[72] Ismay, Bruce J. Testimony, British Wreck Commissioner's Inquiry.

[73] *Survivor Says Ismay Ruled In Titanic's Boats*. The New York Times, Saturday, June 26, 1915. Downloaded on 7/24/2014 from http://www.encyclopedia-titanica.org/says-ismay-ruled-in-titanics-boats.html

that Titanic could operate faster because it had the capability to do so, rather than the strategy to do so, we have a classic demonstration of mistaking technology for quality.

The state of mind, characterized by the arrogant pride of hubris, built on the belief that Titanic was unsinkable, which is evident in Mr. Ismay's testimony, and which served to create a fatally flawed usage scenario for Titanic, in terms of speed, also created at least five additional deviations from quality excellence, in the Process Enrichment Theme of Performance, which significantly affected all of Titanic's stakeholders.

1) Travelling an exceptionally risky course: Testifying at a legal proceeding regarding Titanic's sinking, Captain Henry Meyerdierks reportedly said "if he had received warnings similar to those received by the Titanic he would have gone at least fifty miles south of the danger zone."[74] Titanic's route, although slightly altered, was not altered as significantly as the conditions would dictate in order to avoid the path of the reported ice fields and icebergs.

2) Limiting quality to conformance to requirements: Although Titanic conformed to the requirements of the British Board of Trade in the number of lifeboats it carried, thus meeting the last century's definition of quality, Titanic carried only 20 lifeboats for the 2,201 people on board. Even if every lifeboat was filled completely, there were barely enough lifeboats to rescue half the people on board. Apparently dissatisfied with the conformance to requirements definition of quality, as it was now used to evade responsibility, the Commissioner, in charge of the British Board of Trades' inquiry, questioned Harold Sanderson, Vice President of the White Star Lines parent company, and sought to learn the principle guiding quality in the decision of how many lifeboats to carry.

> "The Commissioner: Now what I want to know is, upon what principle did you proceed, if any, when you fixed the number of boats that you would provide. Why did you stop at the number which you stopped at?

---

[74] *Survivor Says Ismay Ruled In Titanic's Boats.* The New York Times, Saturday, June 26, 1915.

Harold A. Sanderson: I do not think there was anything more definite in our minds than that we should comply with the requirements, and, as is our usual practice, go a little in excess of them, without any particular theory as to what that excess should be."

As we have encountered previously, once conformance to requirements is accepted as the definition of quality, people tend to abandon their responsibility for principled decision making, and perform just well enough to honestly say they did as they were required to do. This is precisely the manner in which the decision of how many lifeboats Titanic should carry was made. Mr. Ismay was presented blueprints by Titanic's designers that included enough lifeboats for all the passengers. In fact, the equipment to hold and deploy 48 lifeboats was already built in to Titanic, however, when the British Board of Trade required only 16 lifeboats, the designers plan was abandoned in preference for only 14 standard lifeboats, 4 smaller collapsible lifeboats, and 2 even smaller cutters. After quality became defined as "a little in excess of" conformance to requirements, quality degraded to the point where the ability to perform a rescue scenario, for all the people in the system, didn't exist – even though some of the world's wealthiest people were on board Titanic and would, probably, be ready, willing, and able to pay for the extra lifeboats.

3) <u>Not practicing safety procedures</u>: With carefully exercised safety drills, simulating real-world emergency scenarios, the crew would have understood better how to fill the lifeboats with passengers. In the event of Titanic's sinking, if lifeboats were properly filled, another 470[75] people would have been saved.

4) <u>Not equipping Lookouts with necessary resources</u>: Lookouts were not equipped with binoculars or night vision equipment which was available and commonly used in 1912. In the testimony of Mr. Frederick Fleet, the lookout who first saw the iceberg which Titanic hit, which he called "a black mass," if

---

[75] Titanic's lifeboats could carry 1,176 people. Since there were only 706 survivors, 470 lifeboat seats were empty.

he had "glasses" he "certainly" could have seen the iceberg "in time for the ship to get out of the way."[76]

5) <u>Not prioritizing wireless messages regarding hazards</u>: Wireless operators, who were aboard via contract with the Marconi Company, preferred to send and receive personal messages for passengers, for a profit, rather than attend to ice warnings from other ships. Several warnings, reportedly, were not delivered to the Captain of the Titanic at all. Upon hearing the very last warning that Titanic would ever receive, from the Californian, sailing nearby, Titanic's wireless operator told the wireless operator of the Californian to "shut up" in order to keep the airwaves quiet so that he could communicate with the Cape Race wireless station to transmit and receive messages for passengers. The Californian's wireless operator listened for a while to Titanic's traffic in personal messages, then, having no response from Titanic on his ice warning, he went to bed. If this response had taken place, it's possible that the Californian's message would have caused the Titanic to slow to a safe speed and, instead of going to sleep, the Californian's wireless operator could have heard Titanic's GQD and S.O.S. distress calls which began less than 20 minutes later. Being the closest ship, the Californian would have been able to reach Titanic in plenty of time to rescue all 2,223 people on board.

As the evidence indicates, on the perfectly clear night of April 14, 1912, as Titanic traveled at "her maximum speed of the voyage - not less than 21 knots, or 24 1/4 miles per hour," which was also the fastest Titanic had ever travelled, Titanic raced through the region of which she had received several ice warnings. The water was so calm that it was difficult to see icebergs because there were not enough waves breaking against them and there was not enough light behind them. For one reason or another, the lookouts were not equipped with telescopes or binoculars, however, at about 11:40 PM, one of Titanic's two lookouts, Mr. Fleet, rang three bells signifying that he saw an "iceberg right ahead." Mr. Robert Hitchens, the Quartermaster, turned the wheel on the order "hard astarboard" given by Mr. Murdock, the First Officer, and attempted to stop and reverse engines in an attempt to evade the iceberg – but it was too late. Titanic crashed into the iceberg, in a glancing blow, at nearly full speed. Many of the passengers

---

[76] Fleet, Frederick. Testimony, British Wreck Commissioner's Inquiry, Titanic Inquiry Project, retrieved on 6/7/2014 from: http://www.titanicinquiry.org/

were undisturbed by the crash as the impact didn't significantly jar the ship. Soon however, Titanic began to sink as water entered through 5 damaged compartments. The crew lowered the electronically controlled watertight doors, which apparently operated correctly, sealing the watertight compartments. Since Titanic's watertight compartments were not truly watertight – they were not sealed on top – as water filled the 5 damaged compartments and weighed Titanic down, the water simply filled one compartment and then flowed over the top and filled the adjacent compartments, like an ice-cube tray, further weighing down and sinking Titanic as water continued to propagate all through the ship. Mr. Ismay boarded the last standard sized lifeboat, as it was being lowered to the water below by dedicated crewmen. Mr. Ismay then helped row away as Titanic sank behind him, taking the lives of 130 first class, 166 second class, and 536 third class passengers and 685 of the crew – more than 1,517 people in all. In the final voyage of the R.M.S. Titanic, it was hubris which caused unnecessary risk in the Process Enrichment Theme of Performance and brought tragedy to all of Titanic's stakeholders.

Epilogue

The story of The Final Voyage of the R.M.S. Titanic exemplifies the precedence of strategy over technology as the basis for quality excellence in creative effort and reveals, with unique clarity, the tragic nature of Titanic's accident. While technology can be a great source of pride, when the level of pride rises to the level of hubris, this can result in a dangerous form of fearlessness that can obscure clear thinking about real-world scenarios in which a product or service will be used – to the extent that dangerous conditions can be ignored – with tragic results.

In creative effort where technology plays a prominent role, it should be realized that technology only attains value when it provides performance with the ability to enact a valued strategy.

Case Study:

The Battle of Britain

The Challenge of the World Class Theme of Performance
___
- How will we achieve performance quality excellence everywhere we perform?

Introduction

The World-class Theme of Performance is the sixth and highest echelon in the hierarchy of quality's six Themes of Performance. The World-class Theme of Performance considers the versatility of a Performance Unit to be used under varying conditions and circumstances. As the expression World-class suggests, this versatility also extends to the ability to enable customers to perform their strategies everywhere, geographically, it is desirable to perform. Addressing the strategies of diverse and dispersed customer groups, on a world-wide scale, requires coordinated use of the knowledge of varying localized ideals of quality excellence as a basis for managing the adaptation of product and service performance to the desires of an extended set of customers. By creating the ability to serve and be attractive to a larger customer base, World-class Performance enables broader benefit to be derived from the successful performance of an organization and an organization's products and services.

World class performance brings to an organization the potential for beneficial alliances and the acceleration of organizational growth and success, however, organizational growth, on a world-wide scale, can also be expected to create rivalries with established local businesses as an organization becomes a new entrant competing in their markets. Beyond mere competition, challenges by rivals should be expected to take everything from rational and legal to, in extreme cases, irrational and illegal forms because entrenched businesses are likely to view a new entrant to their market as a challenge to their organizations' survival. All the defenses available to an entrenched competitor, including political protectionism, better venues for strategic communication, and better and broader relationships with customers, among other defenses, can be used to increase the risk of poor performance for new entrants. Ardent rivals, or "adverse intellects," can be depended on to challenge their minds to conceive of ways to defeat each other's strategies in just about any way they can.

Prologue

The success or failure of world-class performance in competitive warfare and, also, in military warfare depends on the ability of an organization to adapt to dynamic local competitive conditions and to appropriately and effectively apply ingenuity and creativity to respond in expected and, as necessary, in unexpected ways, to competitive challenges.

In the years prior to the outset of World War Two, while the uncomfortable, and, probably, unsustainable Treaty of Versailles, which ended World War One, was in effect, Germany was prohibited by World War One's victorious nations – led by Britain, France, and the United States – from forming an Air Force. The right to ensure themselves peace, was what the victorious nations believed they had paid for with their soldiers' and civilians lives, lost on an unprecedentedly massive scale, and was what the cause of civilization demanded. Secretly, however, Germany laid plans for and began building an Air Force. As the secret got out, nothing effective was done to counteract or control this treaty violation. Instead, the desires to ease the burden of the Treaty of Versailles and to allow Germany to restore her pride, and also to avoid re-opening the hostilities of World War One, influenced the leaders of the victorious nations to tread a path onto the slippery slope of appeasement. In this way, the path of appeasement grew to be acceptable as the victorious powers right and honorable policy. Unfortunately, as slippery slopes tend to do, this path didn't culminate in a contenting plateau, instead, it allowed, and perhaps encouraged, more belligerent behavior and greater treaty violations to occur. For example, as Mr. Churchill wrote, "The acceptance by the Western Democracies of the German subjugation of Austria encouraged Hitler to pursue his designs more sharply against Czechoslovakia."[77] Within Germany, the corrosive effects of being rewarded for belligerence, cheating, and subterfuge exacerbated the worst manifestations of an already dangerously deranged social order. This served to strengthen a political leadership structure which, as Mr. Churchill described it, was "a monstrous tyranny never surpassed in the dark, lamentable catalogue of

---

[77] Churchill, Winston. "The Second World War: The Gathering Storm," (Houghton Mifflin, 1951) p. 251.

human crime."[78] In charge of this "monstrous tyranny," having seized dictatorial powers, was an amphetamine toting "maniac of ferocious genius, the repository and expression of the most virulent hatreds that have ever corroded the human heart,"[79] by the name of Adolf Hitler. Second in command, and commander of the new German air force, was Mr. Hitler's methadone addicted cohort Hermann Göring.

If the policy of appeasement worked as desired, British Prime Minister Neville Chamberlain, following his return from negotiations in Munich, Germany, when he proclaimed that he had obtained "peace for our time," might have been hailed through history as "The savior of Europe." As events unfolded, however, the virtue of Mr. Chamberlain's effort was lost on the likes of Germany's leaders. In consequence, by the critical year 1940, the monstrous tyranny running Germany controlled the world's largest Air Force, with mostly new, state-of-the-art planes and an abundant cadre of pilots and air crews freshly trained in the latest technologies and doctrines for use of air power. In contrast, the governments of Britain and the U.S. led their nations in spending the post-World War One years pursuing the opposite course with their armed services – favoring peace, and a ratcheting down of the instrumentation for war, through disarmament treaties and large scale destruction of their weaponry. For example, on the basis of the treaty signed at the Washington Naval Conference in 1922, the U.S., Britain, and Japan destroyed 17, 19, and 10 active warships, respectively. At the high point of the peace movement, in 1928, many nations signed a "General Treaty for Renunciation of War as an Instrument of National Policy" in an effort to prohibit war. Further progress was made in treaties of alliance against "aggressor nations" in the event they initiated an unjust war. In consequence, although progress towards sustainable world peace was made, world war came anyway, and when world war came, the great democracies were not ready.

Following Germany's invasion of Poland, the inevitability of war was upon Britain and France who were compelled by their treaty obligations to Poland to declare war on Germany since Germany was the aggressor nation. One ramification of Britain's declaration of war, although it came too late to prevent

---

[78] Churchill, Winston. "Blood, Toil, Tears and Sweat" speech in the House of Commons, May 13, 1940
[79] Churchill, Winston. "The Second World War: Their Finest Hour," (Houghton Mifflin, 1951) p. 24.

the defeat of France, in the early stages of World War Two, is that the fog in London began to clear. The apparition of Winston Churchill, rising from the ranks of the bitterest opponents of disarmament[80] and appeasement, to become Britain's Prime Minister, portended the enormity of change in the game plan with which Britain would confront the imminent storm which Mr. Churchill called the Battle of Britain.

The Battle of Britain

The stunningly prompt defeat of the renowned French army, which brought the fall of France, in June, 1940, and worse, the "defection"[81] of France, as Mr. Churchill described it, left vital parts of the British island within range of attack from German air forces based in France - just across the English Channel - an expanse of only 21 miles in its narrows. These air bases would supplement air bases captured from Belgium and Holland which were also being prepared for operations against Britain. Thus, the Battle of Britain developed with a brief lull as a redirection and intensification of the air war from France to Britain occurred.

After rapidly toppling or conquering so many august nations, Germany was in the position of dominating Western Europe. To consolidate this position, Mr. Hitler extended an offer of peace to Britain. This offer was rejected by a British government official without much fanfare. Of this event Mr. Churchill wrote "Hitler could not conceive that Britain would not accept a peace offer after the collapse of France."[82] Realizing that Britain's surrender could not yet be wrested from them, an amphibious cross-channel invasion plan called "Operation Sea Lion" was developed. Operation Sea Lion's top priority was air superiority over the English Channel and landing areas. Although air superiority was aggressively pursued, Hitler hoped that instead of actually invading Britain, which would only be done as a "last resort," a weakening food supply and "terror attacks against English centers of population" - in the guise of "reprisal actions"

---

[80] Mr. Churchill said, in 1932, in the House of Commons "the thesis that they [Germany] should be placed in an equal military position with France is one which, if it ever emerged in fact, would bring us within practical distance of almost measureless calamity." Mr. Churchill's "entourage" included well informed members of the intelligence community such as Desmond Morton and Ralph Wigram who provided evidence to strengthen Churchill's arguments in parliament.

[81] Churchill, Winston. *The Few*, Speech to The House of Commons, August 20, 1940.

[82] Churchill, Winston. "The Second World War: Their Finest Hour," (Houghton Mifflin, 1951) p. 320.

– would have the psychological effect of creating a constant fear of invasion and would cause the people of Britain to be "seized with mass hysteria"[83] and thereby compelled to pressure their government to surrender. Against this hope stood the more fundamental hopes and aspirations of the people of Britain who were still free, however imperfectly, and who could still be emboldened by strategic communication.

In the British parliament's House of Commons, speaking to his countrymen on June 18, 1940, his 40th day in office as Prime Minister, on the occasion of the fall of France, Mr. Churchill eloquently expressed the significance of the Battle of Britain.

> "What General Weygand called the Battle of France is over. I expect that the Battle of Britain is about to begin. Upon this battle depends the survival of Christian civilization. Upon it depends our own British life and the long continuity of our institutions and our Empire. The whole fury and might of the enemy must very soon be turned on us now. Hitler knows that he will have to break us in this island or lose the war. If we can stand up to him, all Europe may be free and the life of the world may move forward into broad, sunlit uplands. But if we fail, then the whole world, including the United States, including all that we have known and cared for, will sink into the abyss of a new Dark Age, made more sinister, and perhaps more protracted, by the lights of perverted science. Let us therefore brace ourselves to our duties, and so bear ourselves that, if the British Empire and its Commonwealth last for a thousand years, men will say, 'This was their finest hour.'"

Winston Churchill
June 18, 1940

From the German perspective, since they possessed the world's largest air force and the world's largest army, expectations for the outcome of the war with Britain must have seemed preordained in their favor. General Alfred Jodl, expressing the view of the German High Command wrote, on June 30, 1940,

---

[83] Hitler quote

that "Germany's final victory also over England is only a question of time... Germany, therefore, can choose a form of warfare which husbands her own strength and avoids risks." Risk avoidance would be a wise approach since this was to be a battle of decisive importance which could not be taken for granted. If Germany won air superiority over the coastal areas and the English Channel, the next phase of Operation Sea Lion would proceed and, assuming the British navy couldn't stop it, Britain would be invaded by an army of overwhelming size[84] and would probably be defeated. Conversely, if Britain repulsed the German air attack, Germany would be unable to control the growth of Britain's air force and, eventually, that air force, backed by the productive effort of Britain, Britain's dominions, allies, and the U.S., would grow to proportions that could lead to Germany's downfall. That this outcome was a considered British goal is evident in a message Mr. Churchill sent to General Smuts on June 9, 1940, in which he wrote "I see only one sure way through now – to wit, that Hitler should attack this country, and in so doing break his air weapon."[85] This belief in the momentous consequences of the intensifying air war, which depended on only a few hundred planes and a few hundred pilots, was epitomized in the classic phrase Mr. Churchill used in a speech just over one month into the Battle of Britain.

> "Never in the field of human conflict was so much owed by so many to so few."[86]
>
> Winston Churchill
> Aug 20, 1940

Although Britain's air force and defenses against air attack had fallen far behind necessary levels of readiness, and Britain's industrial base was also well behind in the capability to manufacture the state-of-the-art aircraft necessary for modern warfare, the traditions, abilities, and lessons of a sophisticated World Class adversary had not been forgotten or lost. From the onset of World War Two, Britain was rapidly rearming to meet the impending threat from the air

---

[84] Memorandum of 30 June 1940, General A. Jodl "We must count with an opposition of about 20 English divisions so that at least 30 German divisions will have to be embarked.";
http://www.loc.gov/rr/frd/Military_Law/pdf/NT_Nazi-Suppl-A.pdf

[85] Their Finest Hour, p. 320

[86] Churchill, Winston. *The Few*, Speech to The House of Commons, August 20, 1940.

from Germany. Although the equipment and infrastructure to support modern warfare became available only slowly, by July 10, 1940, the date regarded as the beginning of the Battle of Britain, each adversary controlled formidable modern air forces and efficient aircraft production and support organizations.

According to Mr. Churchill, "By August, the Luftwaffe [The German air force] had gathered 2,669 operational aircraft, comprising 1015 bombers, 346 dive-bombers, 933 fighters, and 375 heavy fighters"[87] against which, Mr. Churchill reports, the British maintained an average of 608[88] operational fighter aircraft for the duration of the Battle of Britain. These opposing strengths have proven impossible to confirm with great accuracy. There were also hundreds of additional aircraft that were not quite ready for operations on both sides and also reconnaissance and other aircraft types that were involved in roles less crucial to the outcome of the battle. The difficulty in maintaining accurate information on operational readiness, as aircraft were damaged or destroyed or replenished from new production, or drawn in or out of reserve locations, is inherent in a dynamic, fast moving, war. The picture that emerges clearly is of a numerically superior attacking force on the order of approximately 2.15 to 1 in fighter aircraft and 4.1 to 1 in total aircraft. In view of this disproportion of strength, there was no readily obvious reason for the German air force to doubt their leaders' optimistic expectation that Britain would be sufficiently weakened to be invaded by "the end of August or the beginning of September"[89]; however, Germany's leaders would have been well advised to recall the doctrine of the famous German military theorist Carl von Clausewitz.

"Superiority in numbers is the most important factor in the result of a combat, only it must be sufficiently great to be a counterpoise to all the other co-operating circumstances."[90]

Carl von Clausewitz
1832

---

[87] Their Finest Hour, p. 323
[88] Their Finest Hour, p. 717
[89] Jodl, Alfred. Translated memorandum of A. Jodl, Chief of the Wehrmacht, June 30, 1940. Retrieved on 7/28/2014 from http://www.loc.gov/rr/frd/Military_Law/pdf/NT_Nazi-Suppl-A.pdf
[90] Clausewitz, On War, 1832 (Translated from German)

Although superiority in numbers of aircraft provided the German air force a great advantage in at least one of the types of resources necessary to succeed in a battle for air superiority, the essential and decisive resource in any form of warfare is strategy. It is through an adversary's superiority in the quality of their strategy of war that the circumstances necessary to win can be created. Most of the critical elements of strategy that went into the planning on both sides of the Battle of Britain were military secrets of the highest order. For this reason, in 1940, the public was not aware of the new applications of technology and the actions and counteractions, introduced by scientists and other professionals working together with military leaders on both sides of the war. Under these circumstances, world opinion was greatly divided over which side would win the air war in the Battle of Britain. Some influential leaders tried to convince U.S. President Roosevelt not to provide military resources to Britain in the belief that after Britain's army was compelled to retreat from the Battle of France, Britain was already on the brink of defeat; however, Mr. Roosevelt believed differently. Mr. Roosevelt was ultimately vindicated by the results of the Battle of Britain, however, the view that Britain was on the brink of defeat appears quite reasonable had the German strategy of war not been undermined and defeated by counteractions of the British, who applied knowledge of the risks in German strategy to creatively adapt their own strategy to introduce performance that exacerbated their adversaries weaknesses.

The story of the Battle of Britain, as it was known to the world as it unfolded, is in large part a story of the day to day aircraft losses in the air war. News reports of remarkable results claimed by both sides must have captivated the public imagination. The chart of figure 2-7 reveals the progression of losses for both German and British aircraft over the course of the Battle of Britain.

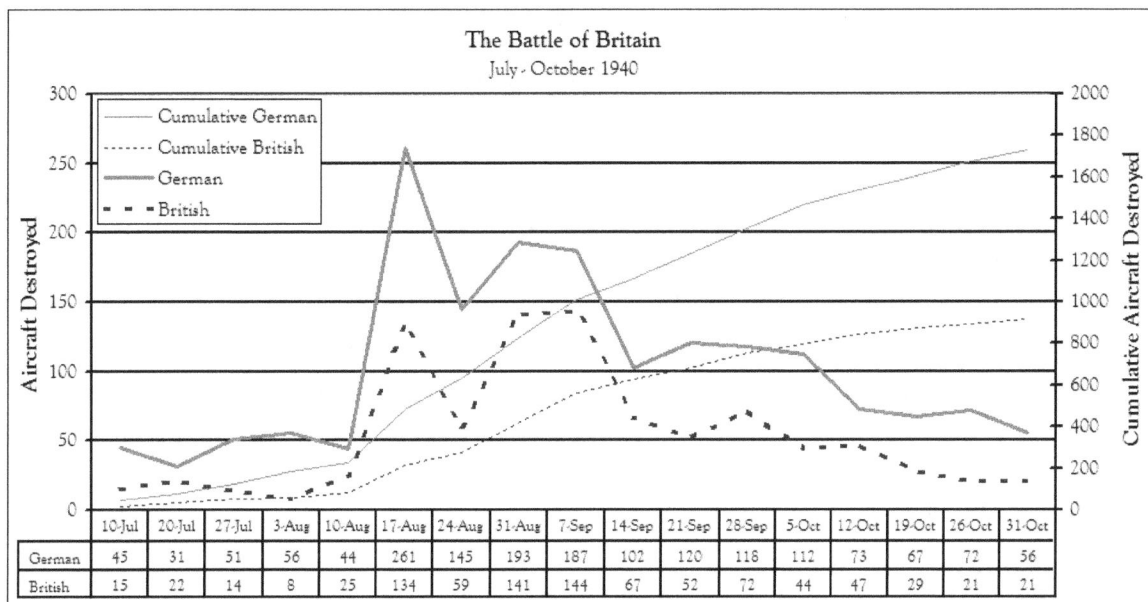

	10-Jul	20-Jul	27-Jul	3-Aug	10-Aug	17-Aug	24-Aug	31-Aug	7-Sep	14-Sep	21-Sep	28-Sep	5-Oct	12-Oct	19-Oct	26-Oct	31-Oct
German	45	31	51	56	44	261	145	193	187	102	120	118	112	73	67	72	56
British	15	22	14	8	25	134	59	141	144	67	52	72	44	47	29	21	21

Figure 2-7; Chart of British and German aircraft destroyed, each week, during the Battle of Britain,[91] showing a cumulative loss of 915 British and 1,733 German aircraft destroyed.

According to secret British records, which survived the war, from the very beginning to the very end of the Battle of Britain, German aircraft losses, shown by the heavy line in figure 2-7, consistently outnumbered British losses by an average of 1.9 to 1. To the public, who were imperfectly informed through news sources and unaware of the secret strategic contest underlying Britain's surprisingly favorable results, this outcome must have been a remarkable surprise. Some of the conditions of the battle were known, and some were predictable, given the advantages of defense. For example, since the British fighters were flying and refueling over Britain, they were able to fly many more sorties than German fighters, on the order of 1.8 to 1, on average, a circumstance which, in effect, nearly doubled Britain's Air Force. During the very heated period in mid-august this ratio of sorties reached as high as 3.5:1[92] – nearly tripling the relative efficacy of the British Air Force. It appears that Britain's production of new fighter aircraft and training of new pilots kept up with losses for the duration of the Battle. The rate of British aircraft production was faster than that of the

---

[91] Their Finest Hour, chart derived from data on pages 339 and 340.

[92] On August 21, 1940, British fighters performed 589 sorties against 170 sorties by fighter aircraft of the German Air Force. Source: The National Archives, Catalogue ref: AIR 2/7771, http://www.nationalarchives.gov.uk

Germans on the order of approximately 1.5:1 and it took the German air force much longer to field replacements and to repair damaged aircraft. The German air force also had to contend with extensive British defenses, including anti-aircraft fire and various contrivances, such as barrage balloons.

It was not long into the Battle of Britain when it was found that the infamous Stuka dive-bomber, with its trademark built-in siren, and precision bombing, which reigned over so much terror and destruction on the battlefields and in the cities of Western Europe, was being shot down in large numbers by the much faster British Fighters – the Hawker Hurricane and the Supermarine Spitfire. Risk was also found in Germany's first line fighters the Messerschmitt Bf 109 and Bf 110 both of which could be outmaneuvered by the slightly slower British Fighters.

Much to Mr. Göring's chagrin, Mr. Churchill's public radio broadcast of September 11, 1940 was rather disparaging. Mr. Churchill proclaimed that "... their losses average three to one in machines and six to one in pilots. This effort of the Germans to secure daylight mastery of the air over England is, of course, the crux of the whole war. So far it has failed conspicuously."[93] Although Mr. Churchill's own postwar statistics showed that the ratio of destroyed planes was closer to two to one, rather than three to one, in Britain's favor – the damage to Germany's reputation was done and Mr. Goring wasn't plausible enough, outside of Germany's own controlled media, to refute the aspersion. This was a risk of poor performance in the German leaders' strategic communication throughout the war and was probably part of the reason that the war was fought through to unconditional surrender rather than to an armistice or a peace treaty which had been the traditional approach to cessation of hostilities in previous wars.

As the air war continued to be costly to Germany in pilots killed and planes destroyed, the German leaders realized that the battle was cutting into the air resources necessary for them to go forward with their other maniacal plans for world domination. This raised in importance their intention to create "mass

---

[93] Churchill, Winston. Sept. 11, 1940 radio broadcast. The figures Mr. Churchill quotes may be significantly exaggerated. The Sydney Morning Herald, of 8/20/1940, put the ratio at 4 to 1 in Britain's favor. Luftwaffe General Stapf, on 8/14/1940, reported the ratio at 5 to 1 in Germany's favor (Source: War Journal of Franz Halder, Vol. IV, p. 160).

hysteria" as an alternative to a prolonged battle of attrition – especially one that they were apparently losing. Rather than continue to bomb strategic targets, the new focus on creating mass hysteria led to the German air force's decision to bomb London, Britain's Capital city, which had a population of more than 8 million civilians. The view of the human drama in the sky which Londoners became familiar with during the "Blitz," as the intense air battle over London was called, is pictured in figure 2-6. Although the intention to bomb civilian targets was an approach approved in the German plans from the beginning, which had not been aggressively exercised, "Hitler had been persuaded by Göring that the major attack on London would be decisive..."[94, 95] British leaders similarly decided that bombing the densely populated cities of Germany was acceptable.

Figure 2-6; Aircraft vapor trails marking furious Dogfights in the skies above the Big Ben clock tower in London, England, during the Battle of Britain.

---

[94] Their Finest Hour, p. 328
[95] At the post-war Nuremburg trial, Mr. Göring defended himself by attributing the decision to bomb Britain's population centers to Hitler.

As the world witnessed the repulse and defeat of the numerically superior German air force, British pilots and the British air defense system must have been regarded with great pride and as a source of hope for Britain's allies. Britain's success cannot be properly accounted for by assuming that Britain had more dedicated or more talented pilots or even better planes. In training, equipment, and fighting spirit this was a closely matched battle – with the numerical superiority of the German air forces neutralized, in part, by the distance they had to fly which reduced the number of sorties they could launch. It is in the closely guarded secrets of the strategies and the technologies driving the behaviors of each adversary where we can reveal the true causes of the superiority of strategy that led to Britain's success.

In the Battle of Brittan, Germany used the emerging technology of radar as a means of accurately navigating aircraft to their targets. The Lorenz navigation system was adopted from Lufthansa airlines where the new system was used in order to improve night flying and flying in rain and fog. For the German air force, this technology was a vast improvement over the traditional system of Celestial navigation, which depended upon a sophisticated method of measurements of star locations. German aircraft navigators or pilots were trained to operate the new radar driven navigation system, which was the world's most accurate navigation system at the time, but the traditional system of celestial navigation – which must have been viewed as obsolete – was no longer taught in air force flight schools. If the system of radar navigation failed, or was made to fail, German pilots would have no reliable, systematic, way to locate their targets. This presented a significant and, as it turned out, a decisive risk of poor performance for the German air force in the World Class Theme of Performance.

The German Lorenz navigation system used a ground station to broadcast a split radio beam along the desired path, which pilots could detect using an on-board radar. The systems operated much like two flashlights shining parallel to each other and overlapping in the middle. The signal in the middle was called the "equisignal" and it was a continual signal, composed of the overlap of the signal on the left which was a long intermittent signal, like a dash, with a short intermittent signal, like a dot, on the right which was the same size as the gap in the dash signal on the left. If a navigator picked up a "dash" signal, he knew he was left of the desired course. If a "dot" signal was detected he knew he was to the right of the desired course. If the equisignal was picked up he knew he was directly on course. German bomber pilots using an advanced version of this system called Knickelbein could navigate within yards of targeted aircraft factories and other strategic targets in Britain in the rain or fog or in the day or night.

Where British scientists came in, and totally undermined this state-of-the-art radio beam navigation system, is in broadcast of their own radio beams with either dots or dashes, which effectively made the German navigators receive an "equisignal" everywhere – telling them they were on course, no matter how far off course they were. The British were also able to broadcast dots or dashes over to one side or the other and thereby "bend" the beam so that German bombers went where British intelligence guided them – for example to empty fields or to fields of decoys where bombs were dropped almost harmlessly. British exploitation of the risks of poor performance in the German radio navigation system and in its improved variants, developed and deployed at great cost and at great difficulty during the Battle of Britain, led to some remarkable incidents. Mr. Churchill wrote that "Certainly one German bomber landed voluntarily in Devonshire thinking it was France."[96][97]

Another risk of poor performance of the "Y" version of the radio navigation beam system was that it included a regimented procedure for signaling to bombers when they were over their target and when they should drop their bombs. In this procedure, a sequence of radio beams were used which told the bombers first when to get ready and then when they were over their target to drop their bombs. British intelligence learned this system and broadcast their own radio beam signaling when to drop the bombs. This worked to the point where not only were German bomber pilots misdirected by the deceptive dash and dot signals, they were also tricked into dropping their bombs too early and on empty fields.

---

[96] Their Finest Hour, p. 339

[97] After the Battle of Britain, according to Mr. Churchill "The bombing of Dublin [Ireland] on the night of May 30, 1941, may well have been an unforeseen consequence of our interference with "Y"." ("Y" was a German navigation system.) Their Finest Hour, p. 389

The circumstance that the German pilots could no longer believe in their radio beam navigation system and had no recourse to celestial navigation meant that bombers couldn't find their targets – a distinct strategic weakness in quality's World-class Theme of Performance. An even greater problem for German bombers was that they were highly vulnerable to British fighter aircraft and they were guided on a course that was literally "broadcast" to British Intelligence. Mr. Churchill wrote "From the settings of the hostile beams and the times at which they played we could forecast the target and the time, route and height of attack." Also, by listening across the English Channel to German planes initializing and testing their radio systems, British Intelligence could count the number of planes that would be attacking. All of this was excellent information to be used by Fighter Command to organize counterattacks, which led to the German planes missing their targets, being shot down in great numbers, and failing to attain their strategic objectives.

During the air war of September and October, 1940, as the German air force recognized that their navigation system was being interfered with, Mr. Churchill recounts that "there is a story that during these two months nobody had the courage to tell Goering that his beams were twisted or jammed." Due to indoctrination that encouraged the belief that the radio navigation beam system was "infallible,"[98] it was allowed to mislead the pilots for too long. The result of conformance to this belief, and the risks which were exploited in their system, was that, according to Mr. Churchill, "the whole German system of bombing was so much disturbed by our counter-measures, added to the normal percentage of error, that not more than one-fifth of their bombs fell within the target areas."[99] Another way to look at this remarkable statistic is that, except for normal error, the attack on Britain could have been 5 times as bad if the British were not able to undermine the German navigation systems. Under this circumstance the resolve of the British people to stay in the war might have been affected quite differently.

---

[98] Their Finest Hour, p. 387
[99] Their Finest Hour, p. 388

126

Epilogue

In the Battle of Britain we can see that it wasn't enough to introduce state of the art technology into the navigation process to make the process have quality excellence. A process must be designed with the capability to perform with strategic quality excellence in order to have quality excellence in any Theme of Performance. The Lorenz system, The Knickelbein system and the newer "X" and "Y" radio beam navigation systems were fine, World-class, systems against an opponent without the intention and capability to defeat them. A state of the art navigation system and superiority in numbers of aircraft, which were undoubtedly considered "best practices" on their own, proved unable to overcome the risk in World-class performance in the German navigation strategy – a condition of risk which the British exploited and turned to their advantage.

> "Our exploitation of the strategic errors which he [Wolfgang Martini, General of the Air Signal Corps[100]] made in the Battle of the Beams diverted enormous numbers of bombs from our cities during a period when all other means of defense either had failed or were in their childhood"[101]
>
> Winston Churchill

In warfare, competition is a rapidly evolving interplay and counterplay of strategy in its most acute and consequential form. In warfare, each adversary competes by attempting to increase the risk of poor performance of their adversary's strategies while simultaneously attempting to preserve, improve and strengthen their own. In the intense conditions of war the dependence on strategy for survival and success make it easy to see strategy as the center of gravity drawing, shaping, and controlling energy, effort, and intellect to its needs. There is symmetry between competition in warfare and competition in commerce which is analogous in all but the extent of destructive consequences. Civilization and humanity introduce restraints on these extreme consequences which enable the vanquished to live and revive to reenter the contest again, perhaps with greater wisdom, perhaps with aspirations towards Coascendancy, and perhaps with a renewed quest for ascendant quality.

---

[100] Military status provided by Wikipedia. http://en.wikipedia.org/wiki/Wolfgang_Martini
[101] Their Finest Hour, page 389

Chapter 3:

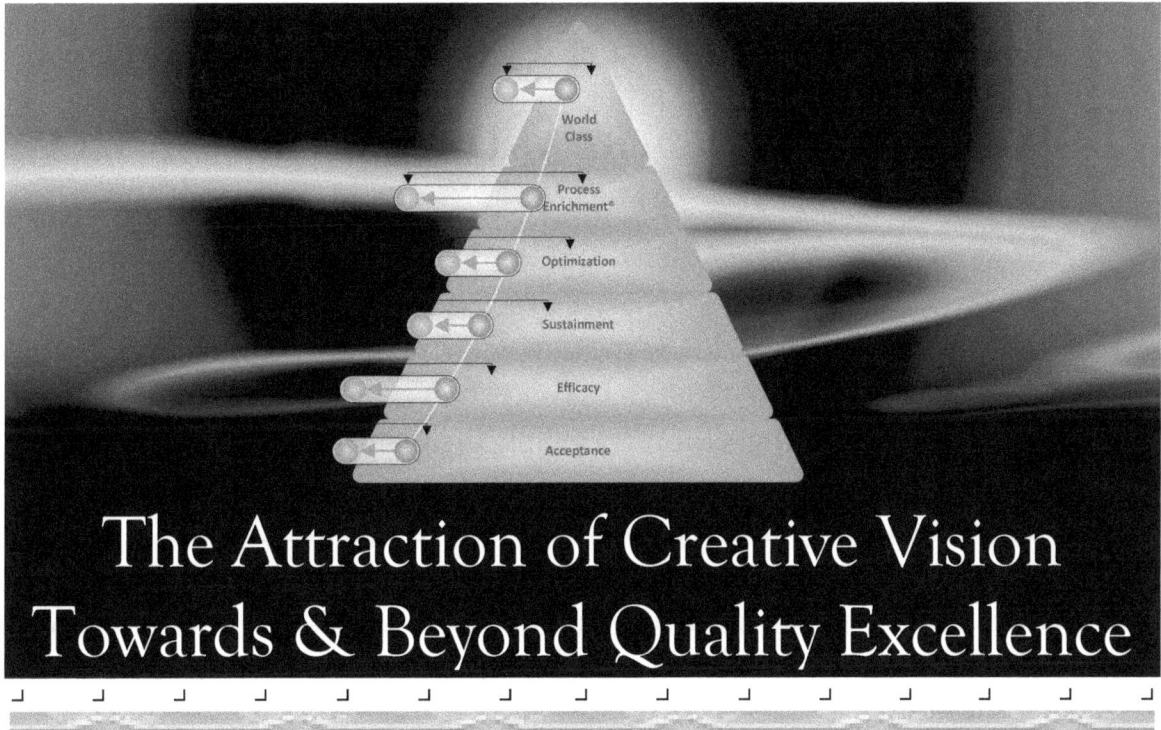

The Attraction of Creative Vision
Towards & Beyond Quality Excellence

*The process of
balancing
quality and risk
is a process common to
all life
in the creation of
all strategies.*

Chapter 3:

The Attraction of Creative Vision Towards & Beyond Quality Excellence

PERSPECTIVES

- ▶ Introduction to the "True Vision" principle highlighting the customer's influence in creative effort
- ▶ The role of historic life-interests in predicting strategic intentions
- ▶ Strategy and its influence as a center of gravity
- ▶ The definition of quality excellence and its role as a standard in creative effort
- ▶ Introduction to the Saile Charting methodology for visual strategic communication
- ▶ Application of Saile Charts to the identification of necessary improvements and to the identification of risks of poor performance in relationship to quality excellence
- ▶ Application of Saile Charts to competitive assessment and to prioritization of improvement effort in order to improve market value over time
- ▶ Transformation to achieve quality excellence
- ▶ The definition of Ascendant Quality
- ▶ The inspired application of creative vision and transformative innovation to achieve ascendant quality
- ▶ Introduction to Strategy Switching and the influences that attract new customers
- ▶ Case Study
  Operation Focus: The Air Battle at the Opening of the 1967 Six Day War
    - ▶ The definition of competitive position

When we experience quality in creative effort that rises to the level of our ideal of quality excellence, we react as if we have experienced something beautiful. Our reaction reflects an emergence of passion because the experiences of quality excellence and of beauty similarly present great appeal to our emotions, desires, and reasoning. Creative effort that appeals to customers' reasoned passions is branded with a very genuine ability to compete and succeed in commerce. The philosopher Plato insightfully highlighted the personal nature of our experience of beauty in a proverb from which we can derive similar insight to quality excellence.

"Beauty is in the eye of the beholder."[102]

Plato
380 B.C.E.

If we are to be the suppliers or acquirers of creative effort that can give rise to the experiences of beauty and of quality excellence, in the eyes of those who benefit from our creative effort, we must have an understanding of the principle driver of customer behavior which, as we have come to know it in the Process Enrichment philosophy, is the customers' quest for ascendant quality. A customer's quest for ascendant quality reflects their personal vision of an improved quality of life and serves as the governance of the strategies they devise to attain it. By empathically understanding the hearts and minds of our customers, and the strategies they devise to serve their quest for ascendant quality, we can perceive the ways in which our creative effort can be attractively formulated to coincide with our customers' vision of beauty and of quality excellence.

Taking the wisdom of Plato's proverb into the fold of the Process Enrichment philosophy we may enjoy a new principle in support of our creative effort.

---

[102] Plato. The ancient proverb "Beauty is in the eye of the beholder" is believed to be derived from the writing of Plato in The Symposium dated 385 to 380 B.C.E.

The True Vision principle can serve to remind us that the quality excellence customers experience in creative effort is a consequence of the ability of the creative effort to enact their most favored strategies in a manner that best serves their quest for ascendant quality.

A supplier's ability to understand a customer's intended strategy, and quality excellence in relation to that strategy, provides the supplier a dependable conceptual basis for the efficient exercise of creative effort that produces products and services that customers will desire and buy. Suppliers who have not managed to conceptualize customer strategy or quality excellence and beauty, in the eyes of their customers, venture forward into the fog of competitive warfare constantly mystified and apprehensive as to the worthiness of their products and services to provide traction and progress or even survivability for their business. This state of mind is reminiscent of, and is perhaps best exemplified by, the state of mind of the British and French public, at the beginning of World War II, as they contemplated the involvement of Russia in the war.

The Russian Enigma

Prior to the outbreak of World War II, Russia sought a firm alliance with Britain and France to counter Germany's aggressive imperialistic expansion – but this effort failed. Consequently, Russia warmed towards Germany and signed a nonaggression pact with them. Soon thereafter, Germany, having reduced their risk of intervention by Russia, began a war against Poland. This aggression compelled Britain and France to declare war on Germany – due to their treaty obligations to Poland – thus initiating World War II. In accordance with a secret provision of the pact between Russia and Germany, Russia joined Germany's war on Poland for the apparent incentive of a portion of Polish territory and resources. There may also have been an incentive to Russia, for the move into Poland, of an advanced front line, closer to Germany, in case a war with

Germany broke out. This move would check the advantage of an advanced front line Germany gained by invading Poland in case they decided to war against Russia. Britain and France weren't obligated by their treaty with Poland to declare war on Russia – and so they didn't – even though Russia was committing roughly the same aggression against Poland as Germany. These circumstances allowed for a great deal of uncertainty and anxiety about Russia's future military intentions. Russia could choose to remain neutral as rival nations destroyed each other in war or Russia could join the war – but it was uncertain on whose side Russia would choose to fight. This choice was uncertain because, on the one hand, Britain and France had now, although late, demonstrated the firmness against Germany's aggressive imperialistic expansion Russia sought before the war – yet, on the other hand, Russia was now enjoying the acquisition of Poland's territory and resources through their alliance with Germany – with the ready prospect of more, relatively easy, acquisitions of other people's wealth.

Although there was great promise of riches in their pact with Germany, there were also great risks. It was well known that Germany's leaders thought little of the Russian people – despite diplomatic words to the contrary. It was also well known that Germany's leaders thought a lot of Russian territory as a target of further imperialistic expansion – despite the nonaggression pact – a pact which Germany's leaders would have diminishing qualms about breaking.

Under these circumstances, one month into the war, Winston Churchill, in the role of Britain's First Lord of the Admiralty, used strategic communication, in a speech broadcast over the radio, in an effort to favorably shape public opinion and, perhaps, to favorably shape events themselves as they were to unfold. In this effort, Mr. Churchill, using his well cultivated oratorical flair, described a viable approach to predicting Russia's strategic intentions.

"I cannot forecast to you the action of Russia. It is a riddle wrapped in a mystery inside an enigma; but perhaps there is a key. That key is Russian national interest. It cannot be in accordance with the interest or the safety of Russia that Germany should plant herself upon the shores of the Black Sea, or that she should overrun the Balkan States and subjugate the Slavonic peoples of Southeastern Europe. That would be contrary to the historic life interests of Russia."[103]

Winston Churchill
1939

Mr. Churchill's insightful message, which expressed a correct presentiment that Russia would eventually side with Britain and France, reveals that Russian strategic intentions could be predictable not only in view of what Russia had to gain in wealth by their alliance with Germany, but also in view of the increasingly unacceptable risks of this alliance. While nations and any other organization of people, like people as individuals, can be depended upon to value and, therefore, to continually pursue, their "historic life-interests" – or, rather, their quest for ascendant quality – it should be recognized that people can always be depended upon to desire to attain their aspirations with the lowest risk of poor performance – and certainly not with an unacceptable risk of poor performance. When confronted with a life threatening risk, people will apply all necessary resources to formulate a strategy to evade or mitigate that risk in order to ensure survival. In this behavior, we are all quite predictable.

As Mr. Churchill's speech has shown, a deep understanding of the elements of strategy, and the well-executed use of strategic communication, served to set the political stage for a Coascendant relationship to form among Britain, Russia, and the U.S, as leaders of the allied powers. This relationship, which came to be known as the Grand Alliance, emerged through a common understanding of a common strategic purpose by the political leadership, the public, and especially the military personnel of the allied nations. It was in support of this relationship that each of the allies shared the capabilities and

---

[103] Churchill, Winston. Speech broadcast on Oct. 1, 1939, quoted in: Winston S. Churchill, The Second World War, Vol. 1, The Gathering Storm, 1948, Houghton Mifflin, p.449

resources they could spare or develop for the other allies to use. This relationship gave rise to the Lend Lease act in which the U.S. provided ships for Britain's war effort, the Western Front, which relieved Russia's fighting on the Eastern Front, and the Russian counteroffensives, and countless other deeds – all of which served the common strategic interest in enabling the allies to work together to win World War II.

An insight which we can derive from this major event in world history is that in the formulation of our strategies our natural inclination to maximize quality while minimizing risk persistently influences our decision making, our intentions, and our behaviors. The interrelatedness among the concepts of strategy, quality, and risk makes them the elements of a predictable and coherent conceptual system rather than isolated conceptual ideas. For this reason, empathically understanding customer strategies, with which customers believe they can best balance quality and risk, in a manner which will improve their quality of life, is key to deciphering the challenging riddle of discovering their needs, unwrapping the mystery of their actions and intentions, and to opening a door to understanding their enigmatic personal vision of beauty and quality excellence. With this understanding, a supplier is well equipped to effectively and efficiently develop capabilities to shape creative effort as close to quality excellence as desired. It is an understanding which bears the potential to enable a true alliance, on the basis of Coascendancy, to form between those who exchange their creative effort for valued incentives and those who acquire and benefit from the creative effort in performance of their valued strategies.

The principle reason that understanding strategy is central to understanding and predicting customer and supplier behavior is that valued strategies organize behavior as though they are centers of gravity around which all the behaviors that make commerce possible are drawn to their purpose. For instance, the politicians' strategy, impelled by the will of the electorate, draws them to their purpose of creating an environment for fair, equitable, and efficient trade. We can also appreciate that customer's strategies draw them to their purpose of selecting and acquiring the creative effort that best meets their needs from among competing suppliers. Suppliers, as we can appreciate, are drawn by their product and service strategy to their purpose of creating competencies to create products and services that customers will find desirable.

The influence that we associate with the "gravitational force" of strategy, compelling us to attain quality excellence in creative effort, and also in the system of commerce, reflects the intensity of the desire for the parties to commerce to attain the incentives of their respective valued strategies with the lowest risk of poor performance. Due to the influence of valued strategies on behavior and on the quality of creative effort, in the Process Enrichment philosophy, a valued strategy is, in effect, analogous to the concept military historians and strategists have referred to as a Center of Gravity (COG). The term "Center of Gravity" is defined and described in U.S. Joint Doctrine.

> "A COG is a source of power that provides moral or physical strength, freedom of action, or will to act. It is what Clausewitz called 'the hub of all power and movement, on which everything depends...the point at which all our energies should be directed.'"
>
> Joint Publication 5-0[104,105]

In military warfare, strategy is the conceptual system governing a course of action necessary to obtain incentives such as peace, independence, or the other nations' resources, etc., that each warring nation desires[106]. In military jargon, these incentives are sometimes called "political purposes" or desirable "end states." The more crucial the success of a strategy is to victory in warfare, the stronger people will be drawn to their purposes of enacting it. One might say, then, that a crucial strategy is a Center of Gravity which attracts military and civilian actions to their purposes. For this reason, the higher up and more broadly influential and critical to success a strategy is, the stronger the gravitational force associated with it will be. The single most crucial strategy may be the one a military strategist would call a nation's center of gravity. By identifying an adversary's strategy as a Center of Gravity, each adversary can then attack every element of this strategy in order to increase its risk of poor performance or to completely prevent its performance. For example, attacks can be made, in a large variety of ways, against the morality of an adversary's

---

[104] "Clausewitz" refers to Carl Philipp Gottfried von Clausewitz, a soldier and military theorist.

[105] Joint Publication 5-0, Joint Operation Planning, 11 August 2011.

[106] This discussion is based on the Process Enrichment philosophy's definition of strategy. Joint Doctrine defines strategy as "A prudent idea or set of ideas for employing the instruments of national power in a synchronized and integrated fashion to achieve theater, national, and/or multinational objectives." Joint Publication 3-0, Joint Operations, 11 August 2011.

incentives, motives, and intentions using strategic communication; obstacles can be raised against certain types of behavior using military forces; and resources could be withheld from an adversary, through an embargo etc. The realization by military strategists of the need to defend their own nations Center of Gravity and attack their adversary's Center of Gravity is reflected in contemporary military doctrine, as exemplified by the idea that "Direct attacks against adversary CoGs [Centers of Gravity] resulting in their neutralization or destruction provide the most direct path to victory."[107] When it is clear that a nations' Center of Gravity is a critical strategy that cannot succeed, that nation is, ultimately, defeated in warfare – unless they can devise and rally around an alternative strategy and keep the war going.

In a customer's effort to perform their critical strategies, the desire to acquire products and services with the ability to perform the services necessary for their critical strategies, at the lowest risk of poor performance, dominates their decision making. This reflects their awareness of the fact that, as in military warfare, weaknesses or risks introduced by inferior products and services can weigh heavily on the ability of their mission to succeed. A customer's proper response to the challenge of defending the success of their critical strategies is to ensure that each element of their strategies and each desired form of performance in each of quality's six Themes of Performance can perform successfully and in a manner that presents a minimum of risk. The customer's defensive intentions, as well as their active desire to attain the incentives of their critical strategy, form the basis of the customer's value system. A customer's value system serves to create attraction towards creative effort with quality excellence and revulsion towards products and services which present risks of poor performance. With regard to strategy, competitive warfare among suppliers is much like military warfare because the relative risk of poor performance for the customer, in the competing strategies offered, determines the winner of the competition.

A supplier who wishes to win in competitive warfare must have superior competence in understanding their customer's or their customer group's critical strategy and in using the gravitational energy of this strategy to draw and shape

---

[107] Joint Publication (JP) 5-0, Joint Operation Planning, 11 August 2011, III-32

their creative effort to the formulation which offers customers a lower risk of poor performance than any competitor. The lowest risk of poor performance occurs at the degree of quality that rises to the level of the customers' ideal of quality excellence in each Theme of Performance. Quality excellence is a point of minimal risk because when a necessary ability to perform is missing, or is present at a capability that is less than quality excellence, poor performance and possibly an inability to successfully enact a valued strategy may follow. It is also true that when the ability to perform is greater than what the customer views as quality excellence the creative effort may have additional features that a customer isn't able to benefit from which may add costs or complexity with counterproductive effects.

Since creative effort is rarely ideal for every customer, and since deviation from quality excellence introduces the risk of poor performance, quality excellence is always achieved by an attainable formulation of creative effort that harmoniously balances a maximum of beneficial performance quality and a minimum of the risk of poor performance. Deviation from quality excellence, other than deviation for the purpose of experimental innovation, should always be recognized as a lapse in the ability to perform a customer's valued strategy – which diminishes the value and competitiveness of creative effort. In an organizational culture where people understand the detrimental effects of deviation from quality excellence, evidence of deviation can be very effective in influencing the effort to create beneficial change.

As our discussion has highlighted, we may understand that "quality is the ability of performance to enact a strategy," however, the concept of quality excellence bears an additional, unique, meaning and purpose that is deeply rooted in the strategy of the person for whom creative effort is intended to serve. These additional challenges for the concept of quality excellence are addressed by the Process Enrichment philosophy's definition of quality excellence, introduced here for the first time.

> *Quality Excellence is the absence of the risk of poor performance, in each Theme of Performance, for a valued strategy.*

Our motivation in the pursuit of valued incentives, such as food or money or anything else we value, is the reason for our development of a strategy. Our strategy improves its value to us as we improve the strategy's capability to attain our desired incentives and as we remove the strategy's risk of poor performance. These improvements in strategy arise as we consider performance in each of quality's six Themes of Performance and, as necessary, substitute methods, behaviors, and resources, which perform our strategy best. If there's a "one best way" of performing our valued strategy, so that we can attain our incentives with a minimal risk of poor performance, we can be counted on to gravitate towards it because it's natural for us to be on a quest for ascendant quality. It is this "one best way" formulation of our valued strategy from which a vision and an ideal of quality excellence emerge.

A customer's valued strategy is the motivating and organizing purpose towards which the creative vision of both a supplier and a customer must converge in order to conceptualize the formulation of creative effort that can achieve quality excellence. When customers evaluate the quality of creative effort, it's the ability of the creative effort to enable them to perform a valued strategy which provides them their true measure of the creative effort's quality excellence. When creative effort possesses the ability to perform, as desired, in each of quality's six Themes of Performance, with an absence of the risk of poor performance, customers will find the creative effort to possess quality excellence, true value, and, perhaps, beauty. When a formulation of creative effort provides the "one best way" of enabling a customer to attain their desired incentives, the creative vision of quality excellence is realized.

The governing influence over the ideal of quality excellence, provided by the customer's valued strategy, reveals that quality excellence is much more than simply "very, very, good quality." Creative effort with abundant, state of the art, technology and ergonomic styling, but absent a strategic purpose that customers demand, is simply very, very, good quality – but not quality excellence. The distinction between various degrees of quality and "quality excellence" is a distinction that we make so easily and so frequently that we can sometimes fail to appreciate our natural skill in making it. As an example, grade school children know to take out a pencil instead of a pen when they go to math class – even though their pen may be their preferred form of writing, with the technology they enjoy. They do this because their usage of a writing instrument in their math class is best served by a simple, low-tech, writing instrument with a capability of erasure. In this example we can see that quality excellence can only exist in the context of a valued strategy. When customers determine the form and degree of quality, in each Theme of Performance, that optimally serves their valued strategies, quality excellence is no longer an abstract concept, rather, quality excellence becomes a creative vision they can clearly articulate – using each element of strategy and each of quality's six Themes of Performance. In this way, the creative vision of quality excellence possesses the unique ability to serve as a standard of performance expressing a customer's aspiration and defining an excellent supplier's measurable goals.

It's instructive to note that strategy defines the intended shape of quality excellence and actual quality defines the actual shape of strategy. For example, if you observe a car that was made under a strategy to serve commuters you will probably find an economical and reliable car. If you observe an economical and reliable car, you will probably infer that the strategy of the supplier was to produce a car for commuters. Although this symmetry between plans and results is the desired and expected outcome of a development and production process, the message of strategic communication can sometimes lose its way as independent decisions are made over the course of product or service development. Sometimes, well intentioned improvements or well-intentioned reversion to what developers thought was working for customers in the past, or even carelessness can act to distort product strategy from its intended shape and increase the strategy's risk of poor performance. The phenomenon of stakeholders adding features excessively is so common in productive organizations that it has its own nick name – "requirements creep." To properly defend any strategy, or to ensure that innovation coincides with the targeted customer group's ideal of quality excellence, proper commitment to strategic communication is essential. In an organizational setting, preserving the symmetry of strategy, and institutionalizing innovative improvement, requires bi-directional strategic communication – from the strategists to the people performing creative effort and back. Clear, concise, easily understood, and professional strategic communication is not only a business essential, it's essential for working with the people in the system who have committed their professional skill to making organizational strategy succeed.

Bi-directional communication of strategic intentions requires a new form of communication that enables the entire organization to create the same mental image of the form of creative effort that achieves quality excellence. Our mental image of the products of our creative effort provides a valuable means for us to manipulate our ideas and form them into something meaningful and useful to us. The Process Enrichment methodology offers us a new form of visual strategic communication which enables us to connect many of the increasingly complex ideas we have been discussing and shape them into an easily communicated visual image. This new methodology, introduced here for the first time, is called Saile Charting.

Saile Charting

Saile Charting is a methodology that is primarily intended to create a form of visual strategic communication which raises awareness of the desired and actual competitive position of creative effort. Using a Saile Chart, a strategist can simultaneously communicate just about every element of strategy, quality, and risk as well as a standard of quality excellence. As we will soon see, Saile Charts resonate with the wisdom of the adage "a picture is worth a thousand words." Although the greatest advantage of Saile Charts is that they enable everyone to know what quality excellence means for their creative effort, as we will soon see, there are a wide variety of ways to use Saile Charting to great advantage.

Typically, a Saile Chart is composed as a qualitative assessment, blending a descriptive sample of information from a variety of qualitative and quantitative performance assessments. Although the information supporting a Saile Chart can be as detailed as the presenter wants, by using a descriptive sample, we are including information that is meaningful and characteristic of the entire issue of concern. In the information contained in a descriptive sample, just like the flavor in a slice of rich cake, we have just enough cake to enjoy so that we also know what the rest of the cake tastes like – without having too much on our plate. Descriptive sampling is a practice that is commonly used in statistics, simulation, and in other mathematical modeling because it enables people to understand the "big picture" with a level of clarity that enables well informed, although not totally informed, decision making to take place. Stakeholders, or those who have an interest in the outcome of decisions being made, can determine if a concern merits further investigation. For this reason, construction of a Saile Chart requires that strategists put just enough cake on stakeholders' plates so that they can comfortably digest it.

The realization that at the heart of management are impressions and opinion guided by the synthesis of imperfect information is exemplified by the planning and subsequent operational excellence on the Israeli side of the, 1967, Six Day War. In the Six Day War, Israel launched a devastating pre-emptive airstrike against the air bases of the aggressive neighboring states as they prepared for attack. Describing the planning of the airstrike, Mordechai Hod, who commanded Israel's air force during the war, reportedly said:

> "Moked was presented on one page to the general staff and without too much detail."[108, 109]

The result of this surprise air attack "presented on one page" was the destruction of virtually the entire air power of the opposing nations within two days – with low losses to Israel's air force. In this example, we can see that high quality information rather than a high volume of information can be very effective. One could imagine that the plan for Operation Moked didn't waste much space with statistical process control charts or with other mathematical information other than how many of *them* are against how many of *us* – simple math.

---

[108] Abraham Rabinovich, The War Nobody Wanted, 6/13/1997, accessed on 5/19/2014, http://archive.today/7U4Jd
[109] The Case Study – "Operation Focus: The Air Battle at the Opening of the 1967 Six Day War" is presented at the end of this chapter.

In the Process Enrichment philosophy, simple mathematics is often more effective than elaborate mathematical techniques. As we will explore in greater detail, quality can be assessed or transformed to quality excellence more easily by using simple addition, subtraction, multiplication and division, and the calculation of percentages, than it can be by using advanced math. The reason this may be surprising is that in the Process Enrichment philosophy we focus on *deviation* from quality excellence, as a basis for quality improvement, whereas in the quality improvement philosophies of the last century the focus was on control of *variation* from quality targets. For example, in the Process Enrichment philosophy the quality of a coat hanger depends, in large part, on the answer to simple, overarching, questions in each Theme of Performance such as "Can it hold a coat?"; "Can it hold the coat's shape?", etc. We can answer these questions "yes" or "no" as in binary math which is simple to use yet powerful enough to be the basis for the logic of today's fastest computers. However, in the last century, when manufacturing was prominent, the quality of a coat hanger was determined, in large part, by how well the diameter of the coat hanger wire conformed to requirements. This determination was usually handled using statistical techniques which had a great deal of focus on quality control and little, if any, focus on quality for customers. As you may confirm, the products and services of today, in our country, are tremendously different and more dependent on creative ability than they were in the last century. This is why, now, and in the future, we will need talent in understanding and interpreting qualitative and quantitative information about creative effort and how this information reflects the relationship of actual performance quality to performance quality excellence. This is a talent that uses simple questions, simple math, and Saile Charts.

In a nation whose revolutionary creed, as written in our Declaration of Independence, begins with the phrase "We hold these truths to be self evident" we must, once again, find ourselves able to formulate and respect each others' qualitative opinions – or we will find ourselves, as a society, confusing our way out of ascendant quality by fruitlessly searching for facts – statistical or otherwise – which will almost always be insufficient for real-world decision making.

Saile Charts are intended to be informative, yet easy enough to understand so they can be influential in driving decision makers to make necessary changes under any circumstances where strategy, quality, and risk are a concern. Saile

Charting, like the concept of quality itself, can be applied to any idea, product, service, project or organization, etc. and, also, to any sub-part of any idea, product or service, etc. This broad applicability is one benefit, among many, of a definition of quality that means the same thing under any circumstances or applications.

Our first use of a Saile Chart will be to visually compare quality excellence, as defined by a targeted customer group, to the actual level of quality of a product or service that a supplier produced for that customer group. As pictured in the Saile Chart on the left side of Figure 3-1, we can see an Areteve[110] upon which two curved lines have been drawn, through two sets of points. The green line, on the left side, labeled "Performance Quality Excellence," represents a strategist's estimate of the quality rating of the product or service, in each Theme of Performance, that the predominant customer group is ready, willing, and able to pay for and that provides a reasonable profit to the producer. The blue line, on the right side, labeled "Actual Performance Quality," represents a strategist's estimate of the actual quality rating for the product or service in each Theme of Performance. Note that the direction of better quality is to the left.

---

[110] The Areteve is a pyramid consisting of six hierarchical echelons symbolizing quality's six Themes of Performance. The Areteve is used for contemplation and evaluation of strategy and the balancing of quality and risk. The name Areteve is derived from the Greek term Arete used to describe excellence and also from the French name Arête given to a wedge shaped mountain peak created by the melting of glaciers, on both sides of the peak, over the course of centuries.

"It is no use saying, 'We are doing our best.'
You have got to succeed in doing what is necessary."

Winston Churchill

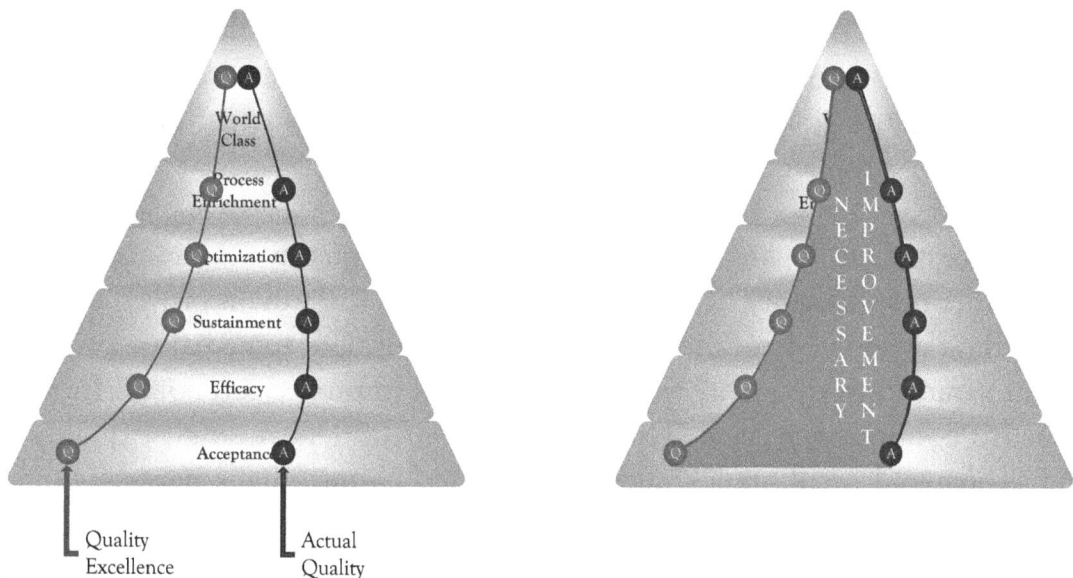

Figure 3-1: A pair of Saile Charts illustrating the necessary improvement required to transform actual performance quality to performance quality excellence.

As the Saile Chart of Figure 3-1 illustrates, the predominant customer group is ready, willing, and able to pay for a product or service with performance quality rated much better than the actual performance quality offered. Although the supplier may feel content with current profitability, based on this extreme difference between what is desired and what is offered, the supplier, in reviewing this Saile Chart, should realize that improvements are necessary and that making them would be profitable. This insight is shown in the Saile Chart, on the right side of Figure 3-1, by the shaded area labeled "Necessary Improvement." Enacting the necessary improvements, which would bring the Actual Performance Quality line over to the line representing Performance Quality Excellence, would also have the very desirable effect of improving the quality of life for customers by providing them products and services that are decidedly better able to perform their intended strategies as they desire. For this reason, if the necessary improvements are enacted, these improvements may work, over time, to significantly improve the supplier's repeat business, market share, and profitability.

What we can enjoy about the depiction of the space between Performance Quality Excellence and Actual Performance Quality as "necessary improvement" is that the word "necessary" is a rather forceful call to action. It's an influence towards action to improve that arose in consequence of communicating and understanding the gravitational energy of the targeted customer group's strategy as it affects the supplier's creative effort.

Once people in an organization understand quality excellence, and that improvement is necessary, it becomes an influence towards improvement which can become a resolved commitment to take appropriate remedial or progressive action to transform to quality excellence, or to an improved level of performance, to ensure continued success. This is the type of understanding of quality that helps evolve and inculcate a strategically focused culture that's necessary for any organization. Besides enabling the creation of great products and services, the ability to respond to the influence of customer strategy makes an organization a great place to work by enabling employees to take part in the creative effort of producing products and services that customers will buy and enjoy. The pursuit of quality excellence also enables employees to sharpen their skills and best use those skills in everything they do. The ability to use quality excellence as a standard also tends to prevent undesired technology additions, requirements creep, and other forms of wasted effort and wasteful use of resources that are symptomatic of organizations that have a weak capability to understand and communicate what quality excellence is.

While the attractive influence of a customer's valued strategy is a compelling reason to make necessary quality improvements, a supplier of creative effort may recognize the additional influences of their own product or service strategy that may also cause the standard of quality excellence to be an ideal state of performance. This additional influence towards quality excellence can be appreciated through analysis of the pair of Saile Charts illustrated in Figure 3-2.

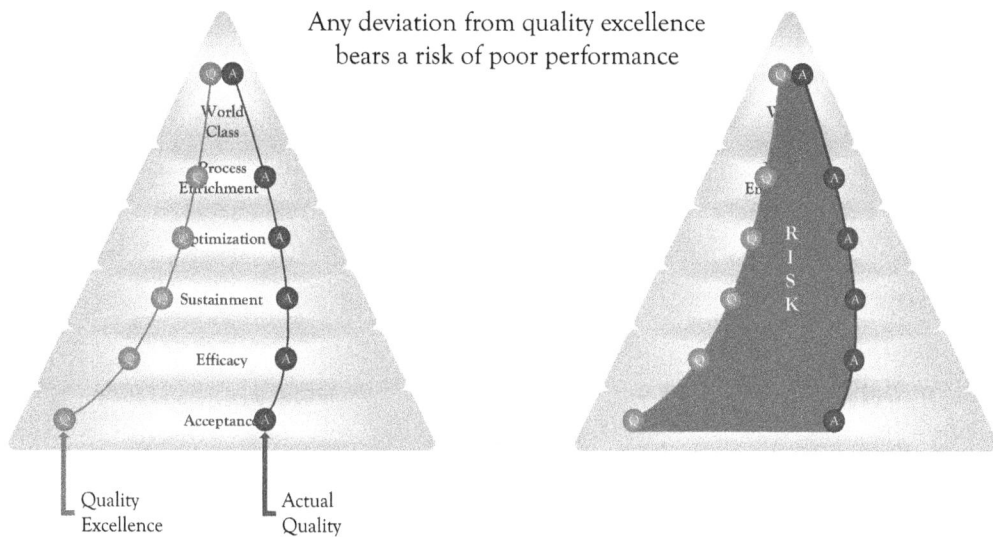

Figure 3-2: A pair of Saile Charts illustrating risk, in each Theme of Performance, caused by deviation from quality excellence.

As shown in Figure 3-2, the void in performance quality in the zone between the green "Performance Quality Excellence" line on the left and the "Actual Performance Quality" line on the right, depicts the risk of poor performance.[111] Risk is a deviation from quality excellence that presents a poor capability or an incapability to perform some aspect of a customer's valued strategy. Risk is also a form of competitive poor performance because poor performance in the ability to perform a customer's strategy creates an opportunity and an incentive for competitors to develop better products or services and thereby capture market share and erode a supplier's profitability.

Analysis of Figure 3-2 reveals that the performance quality rating of product or service (A) can be profitably improved all the way up to (Q) in each Theme of Performance. This may involve improvements in technology and service as the customer desires. The green line, as the standard of quality excellence, is an estimate of the highest performance quality before demand for better performance quality drops off sharply. The performance gaps, between (A) and (Q), in each Theme of Performance, represent an opportunity to reduce the risk of poor performance while increasing performance quality and value. Improvements in this zone may also serve to increase the number of customers in

---

[111] Risk is a concept that we define and explore in great detail in chapter 4.

the market for the product or service. Since customers are willing to pay for the improvements from (A) to (Q), these improvements are almost always necessary.

As we can see, in Figure 3-2, a supplier who desires to attain the profit or other incentives of their own strategy, will find that deviation from quality excellence, as defined in accordance with the customer's strategy, bears a risk of poor performance.

In a Saile Chart, we can identify risk or necessary improvement very specifically in the Themes of Performance which are affected – which is information that has immediate significance. Risk of poor performance in the Acceptance Theme of Performance indicates that a customer finds a product or service unacceptable, to some extent, and, for this reason, will only buy it under stress and possibly only with a significant discount. Poor performance in the Efficacy Theme may indicate a lack of effectiveness that a customer may have to compensate for by acquiring additional products and services. Poor performance in the Sustainment Theme may indicate that a product is more disposable and therefore probably more costly and difficult to manage than the customer would like. Poor performance in the Optimization Theme means that the customer may have to alter their strategy for use of the product or service or be content with a reduced ability to serve some of their needs. Poor performance in the Process Enrichment Theme indicates that a product and service aren't adequately designed for the needs of the people in the system to the extent that the customer may be unmotivated or uncomfortable using a product and service and the supplier may be unmotivated or uncomfortable creating, producing, or delivering the products and services. Poor performance in the World-class Theme indicates that the customer may find that the product or service lacks the flexibility to be used in the variety of conditions in the dispersed parts of the enterprise where its use is desired. Saile Charts are also useful in presenting us with a visual indication of how much we should improve and when we have improved enough to achieve quality excellence. This visual indication is supported by the strategists' qualitative and quantitative assessments.

While strategy serves as the gravitational force, tending to draw the formulation of creative effort ever nearer to quality excellence, we should realize that quality excellence is a concept that is defined within each of quality's six

148

Themes of Performance. The Process Enrichment philosophy has adopted the useful term "Decisive Point" from the military services to describe each of the six points describing quality excellence in each Theme of Performance. The term "Decisive Point" is defined and described in U.S. Joint Doctrine.

> "A decisive point is a geographic place, specific key event, critical factor, or function that, when acted upon, allows a commander to gain a marked advantage over an adversary or contributes materially to achieving success (e.g., creating a desired effect, achieving an objective)."[112]

Joint Publication (JP) 5-0

Achieving a marked advantage over adversaries in warfare or over competitors in commerce requires the astute leader to act upon an understanding of their strengths, weaknesses, opportunities, and threats as they exist at each Decisive Point in each Theme of Performance. When the ability of creative effort to perform a valued strategy achieves the ideal of quality excellence, Decisive Points are in their strongest and most desired state. When quality excellence hasn't been achieved, the Decisive Points of competing products or services may be at a level of quality well below quality excellence. The deviation from quality excellence of competing products and services, in each Theme of Performance, is the

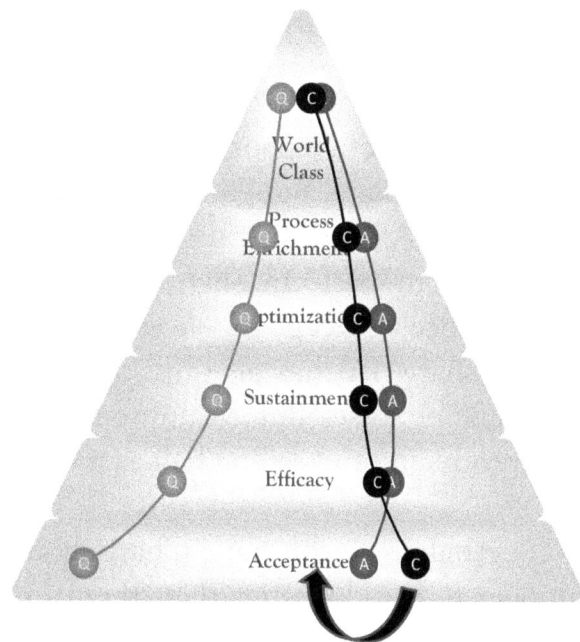

Figure 3-3: A competitive battle with a contested Decisive Point in the Acceptance Theme of Performance.

---

[112] Joint Publication (JP) 5-0, Joint Operation Planning, 11 August 2011, xxii

basis for customers to decide who wins the competition for their business. For this reason, the position of Decisive Points, within each Theme of Performance, provides deep insight to the nature of any competitive battle.

In the competitive battle, pictured in Figure 3-3, we have a Saile Chart consisting of three lines through three sets of Decisive Points in each Theme of Performance. The green line depicts a strategist's estimate of a targeted customer group's ideal of quality excellence for a product or service. The blue line depicts a strategist's estimate of the actual quality, in each Theme of Performance, of supplier (A)'s product or service, which has 50% of the market share. The yellow line depicts the new, competing, product or service of supplier (C), which has gained the other 50% of the market share.

As our Saile Chart, in Figure 3-3, illustrates, the poor performance of supplier (A)'s product or service was an incentive for supplier (C) to enter the market. The cost of poor performance for supplier (A), so far, has been 50% of their market share. Supplier (C) was able to exploit the risk of poor performance of supplier (A)'s product or service because supplier (C)'s product or service has been able to perform slightly better than supplier (A)'s in each Theme of Performance – except the Acceptance Theme. With (A's) loss of 50% of their market share, (A) has also lost approximately half of their revenue and is also experiencing losses due to unused productive capacity since the organization is now producing and selling half as much.

Analysis of this Saile Chart should reveal to supplier (A) the potential for supplier (C) to do better than them in the Acceptance Theme of Performance. Since performance in the Acceptance Theme is the only remaining Theme supporting (A)'s critical strategy to be a supplier of these products or services, this risk may cost supplier (A) the complete loss of their market share. Supplier (A)'s recognition of their risks of poor performance and their attention to necessary improvements is long overdue. For supplier (C), if the value of the additional 50% of the market share is greater than the investment necessary to exceed the performance of supplier (A)'s products and services in the Acceptance Theme of Performance – they should make the necessary improvements. With these improvements, supplier (C) should expect to capture the entire market for their product or service. Supplier (C) should plan to use some of their new profit to invest in the capability to improve up to (Q), in each Theme of Performance, to discourage new competitors from entering the market. Otherwise, new competitors will exploit supplier (C)'s risks and defeat them the same way they defeated supplier (A).

The Process Enrichment philosophy's assertion that strategy, quality, and risk are integral parts of a conceptual system is the basis for Saile Charting. As we have already experienced, Saile Charts present a disciplined way for you to visually organize the elements of strategy, quality, and risk so you can create a useful mental image of them and so you can see how the elements of these concepts may trade off amongst each other. You may find, as I have, that if you remember that a strategy is used to obtain an incentive and that there are six Themes of performance, you can create a mental image of the quality of any product or service and use that mental image to communicate quality interests with great thoroughness. Also, as a discipline for listening and learning, Saile Charts enhance your talent in receiving information from customers or senior managers in a very professional way so that you can picture the elements of quality and the expectations for quality excellence that are being communicated to you. As you gain a deep understanding of your customer's strategy and performance quality desires you can ask focused questions to uncover details that may have been overlooked. Creating a mental image of quality is certainly an exercise that's worth trying anytime you are called upon to give or receive a briefing of any kind or develop or acquire a product or service of any kind.

A brief overview of the key elements of a Saile Chart, as used in the Ascendant Strategy model, is provided in Figure 3-4. The integration of military concepts and terminology reflects their enduring relevance to our effort to communicate about performance quality through the unique perspective of the Process Enrichment philosophy.

"A new type of thinking is essential if mankind is to survive and move toward higher levels."

Albert Einstein

Decisive Points – A Decisive Point is a critical part of the battlefront where an advance or loss in quality could be decisive on the outcome of a competitive battle.

Battlefront – The Battlefront is a line through the Decisive Points in each Theme of Performance. The battlefront holds the shape of the product and service strategy and it establishes the competitive position.

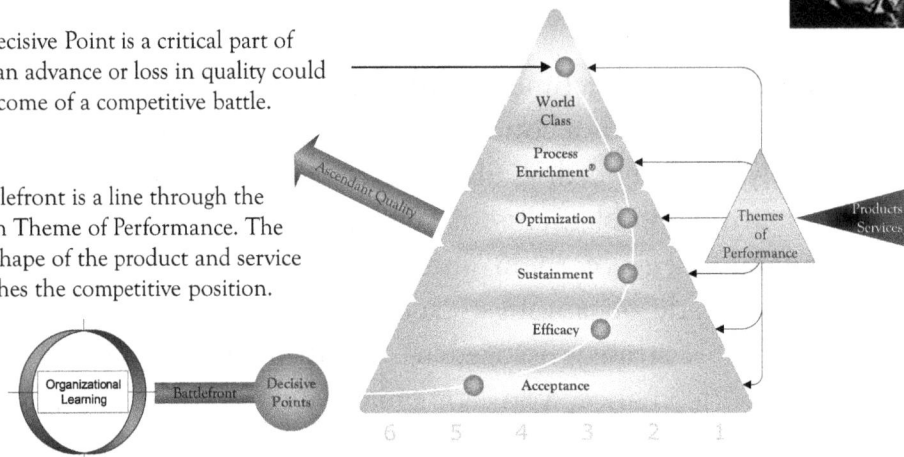

Scoring – An aggregate rating, from 1 to 6, of actual or expected performance, in each Theme of Performance, is shown by the position of each Decisive Point. A rating of 6 equates to quality excellence. A rating below 1 indicates unacceptable performance.

Figure 3-4: A blueprint for visual strategic communication using Saile Charting

As we have already experienced, in Saile Charting, Decisive Points can be positioned, in each Theme of Performance, to represent a strategist's estimate of quality excellence for a product or services' ability to perform a customer's valued strategy. Since quality excellence represents the desired level of quality that a customer is fighting for in commerce, it is appropriate to call the line through the Decisive Points indicating quality excellence the customer's battlefront.

From a supplier's perspective, their fight is to produce a competitive level of quality in each Theme of Performance and an overall strategic offering that surpasses competing alternatives and is attractive to customers. For these reasons, we can say that a supplier also has a battlefront that represents their strategy because the positioning of the Decisive Points, which is based on an estimate of the actual quality of a supplier's creative effort, is a direct indicator of how the supplier wants to compete – or, at any rate, is actually competing. As we have seen, competitive battles are determined directly from the ability of products and services to perform on the battlefront.

In Figure 3-4 we introduce a scoring system which allows for a more quantitative comparison between different Battlefronts. As the topic of risk is developed in the next chapter, we will discuss a simple quantitative method for evaluating and comparing strategies in terms of a Thematic Risk Assessment. Another feature of Figure 3-4, is that, since quality can be improved infinitely, the scale of the quality scoring runs from zero, on the right, to infinity on the left – the direction of improvement towards Ascendant Quality. A score of 6 indicates a state-of-the-art ability to perform. A score greater than 6 indicates an ability to perform bearing transformative innovation and ascendant quality to a significant extent.

Saile charts will be found useful in visually communicating the planned transformation of a battlefront, for a product and service, over time, from a current set of capabilities to a desired future set of capabilities. In Figure 3-5 we see an example of a series of Saile Charts, with transforming battlefronts, visually describing progressive improvements in the ability to perform that some product or service will gain on its journey to quality excellence. Our example product or service has current performance – shown in the Saile Chart on the left – which a strategist has determined to be significantly less than quality excellence in each Theme of Performance. The deviation from performance quality excellence is displayed as a gap in performance between the green Performance Quality Excellence line and the blue Actual Performance Quality line. This assessment indicates that the product or service is probably an immature early model or a first attempt at a challenging type of service. This can also be a product or service that hasn't kept pace with emergent customer needs in the marketplace and which now has critical vulnerabilities or risks that are a competitive disadvantage. Recognition of a more profitable market for the product or service, if it is improved to provide specific types of value to customers, which would reduce the risk of poor performance, has, rightfully, convinced senior managers, engineers, and designers to recognize that several necessary improvements must be made. Since the necessary improvements are extensive, and since the product or service is already being sold profitably, it is decided to proceed with improvement in two phases.

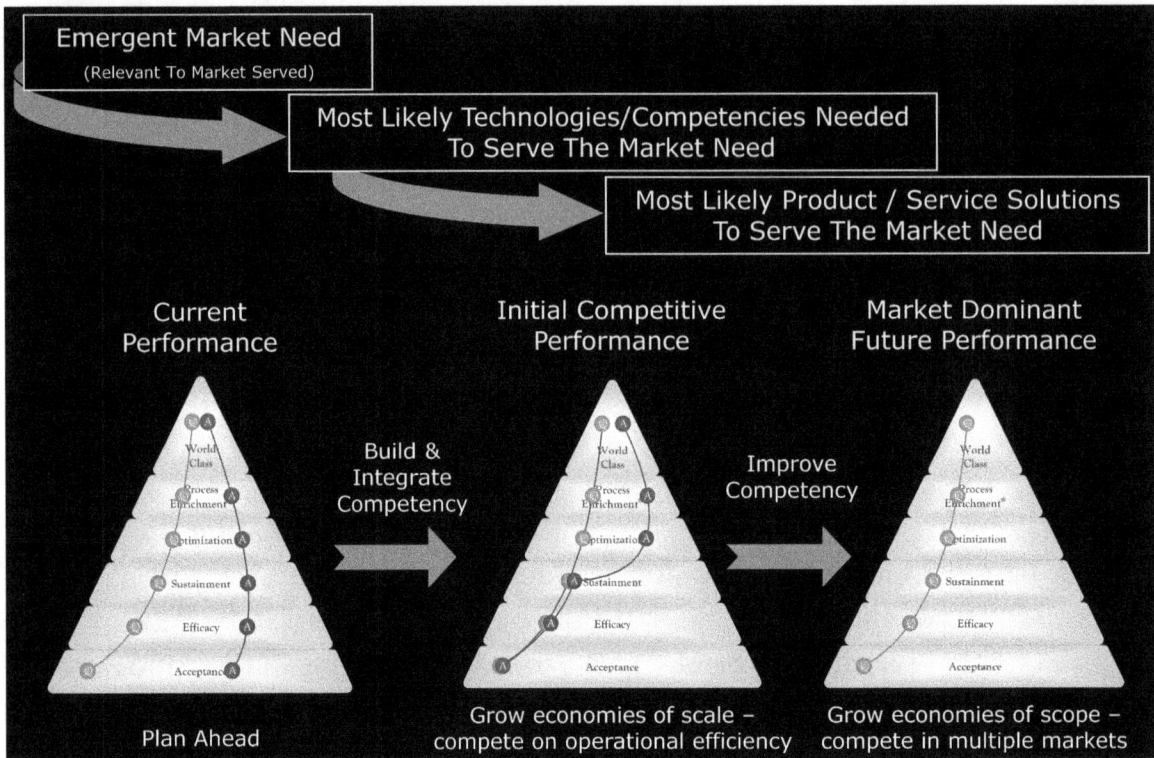

Figure 3-5: A supplier's planned and controlled process of transformation to achieve quality excellence, in each Theme of Performance, in response to needs emerging in the targeted customer market.

When improvement is rolled out in phases, priority should be given to making improvements in the Theme or Themes of Performance in which the most profit would be gained, for the amount invested, in order for improved profits to help fund subsequent phases of product and service improvement. In this example, improvements will begin in the Acceptance, Efficacy, and Sustainment Themes of Performance because these improvements will make the product or service acceptable to a larger number of customers, more effective in performance of their strategies, and will also offer more reliable and sustained service. According to market analysis, these improvements will increase the value and desirability of the product or service to a larger number of potential customers and should improve revenue and profit. When organizational capabilities mature sufficiently to enable efficient creation of quality excellence, in the first three echelons of the Themes of Performance, the Initial Competitive Position, as shown in the Saile Chart in the middle of Figure 3-5, will have been achieved. Products and services that are stronger in the lower echelon Themes of

Performance are likely to compete on things like price and basic ability to perform. Commodities like chemicals, gasoline, and many types of food and also services that are very common in the marketplace like inexpensive motels, dry cleaners, and hamburger restaurants will often exhibit the shape of performance quality shown in the Saile Chart in the center of Figure 3-5.

Since our example product or service isn't a commodity, and customers demand more optimal performance that is easier to use wherever they use it, the supplier is, of necessity, progressing further towards quality excellence in the higher echelon Themes of Performance – as shown in the third Saile Chart of Figure 3-5. Completion of this second phase of progress has enabled the supplier to create a product or service which is competitive, in multiple markets, on the firm basis of performance quality excellence – which customers are likely to find attractive and desirable, and at a price they are ready, willing, and able to pay.

In each phase of progress, as a product and service are transformed to an improved strategic battlefront, where they can compete best, the Saile Chart provides a very realistic picture of a strategist's assessment of progress as the basis for strategic communication as well as for measuring, monitoring, and controlling progress in the transformative creative effort.

The Inspiration Beyond Quality Excellence

As individuals or as part of an organization, we shouldn't consider attainment of our vision of quality excellence to be an end-state of our creative effort, rather, we should be inspired by our achievement of quality excellence to venture beyond our current ideal to conceive of new and better ways to attain our desired incentives and to help our customers attain theirs. Knowledge of our customer's quest for ascendant quality and the incentives for which customer strategies are designed provide the context and insight necessary to predict the nature of desirable forms of quality beyond the state of the art. On this basis, knowledge of quality excellence can be a foundation upon which an innovative organization can best conceptualize goals for research and development and for the integration of technology in their products and services that can transform and improve customer strategy to enable customers to more easily attain their desired incentives.

Organizations willing to venture beyond quality excellence to speculate on innovative new products and services, which have the potential to better serve their customers' quest for ascendant quality, are those rare organizations that inspire and enchant their customers and that are, deservedly, admired by virtue of their great ability to serve as the engines of creative improvisation and progressive momentum in our society. The characteristic response to organizational ascendant quality and ascendant quality in products and services is a stream of demand from a great tide of customers who are often willing to speculate, in good faith, on the suitability of the new, innovative, offerings of these organizations to serve their aspirations. In this way, speculative risk in creating ascendant quality may serve to create a virtuous cycle of Coascendancy in business relationships and, also, a virtuous cycle of social progress.

The pursuit of new and progressively higher levels of ability to perform the strategies which improve our quality of life has been and will be mankind's enduring quest because even quality excellence can become a level of quality that creates an unsustainable arrested culture as, over time, the necessity and desire to improve influences and inspires the creative imagination. The influence of the aspiration to improve beyond what has previously been accepted as quality excellence is the source of the compelling energy creating centers of gravity around improved strategies which emerge over time. From new strategies which emerge to improve attainment of incentives, or to better attain combinations of incentives, new ideals of quality excellence also emerge and, in commerce, the capabilities necessary to produce the new ideal of quality excellence evolve to attain this new standard. In this way, Ascendant Quality is a product of the transformational influences of aspiration, inspiration, invention and innovation in creative effort. The Process Enrichment philosophy provides the definition of Ascendant Quality, which is introduced here for the first time.

156

While quality excellence serves to ensure achievement of customer satisfaction and competitiveness, and may enable an organization to achieve market leadership, in contrast, Ascendant Quality's primary purpose is to drive creative effort to a desirable competitive position that creates market leadership by redefining quality excellence for the customer market at a new and higher level of performance. Success in the speculative effort of leading the market as a merchant of Ascendant Quality is dependent not only on the organizations' ability to develop or absorb emerging technology, integrate it and transform their products and services for the better – but is also dependent on the customers ability to recognize improvements that are offered which help them attain their desired incentives easier and in a desirable manner. The critical need to educate customers on the benefits of innovative products and services compels innovative companies to pair their innovative competencies with strong competencies in strategic communication. In this way, through strategic communication, the innovative organization can improve the extent of favorable competitive information in the market, propagate the virtuous cycle of Coascendancy to a broader base of customers, achieve market success, and improve their contribution to the progressive momentum of society.

A conceptual view of the introduction of market transforming Ascendant Quality, to a product or service which has already attained quality excellence, is illustrated in Figure 3-6. The transformation to Ascendant Quality is, ideally, funded through the market success of quality excellence.

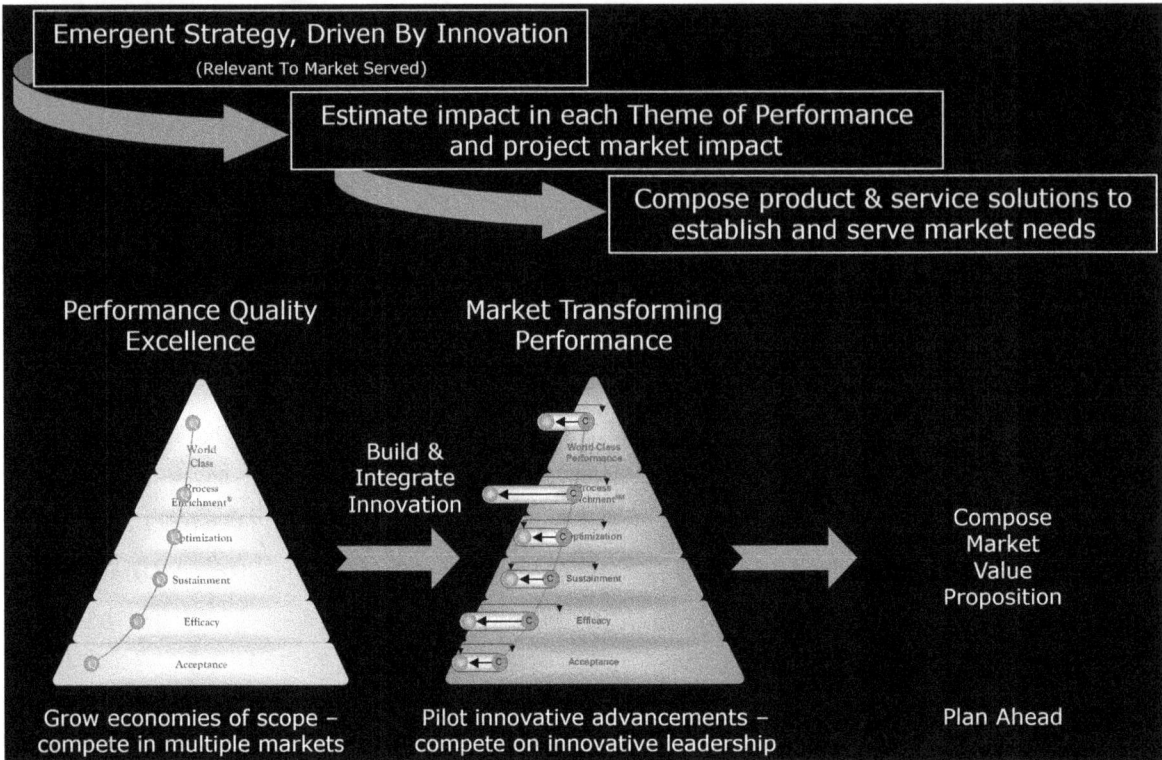

Figure 3-6: Applying innovation and strategic communication to create ascendant quality and transform market demand to a higher and previously unknown level of quality.

From the customer's vantage point, the ability to assess the relative quality excellence of alternative products and services available in the market is an essential talent required for reducing the risk of poor performance in procurement decisions. The customer's ability to engage Coascendant relationships with suppliers further reduces their risk of poor performance and offers them the potential to more regularly attain ascendant quality and accelerate successful performance for themselves and for their suppliers. This centrality to commerce of the weighing of quality, by both sides of marketplace transactions, is the main influence creating a tendency towards ascendant quality of products and services, over time, as suppliers bring more suitable and valuable products to the market and as astute customers build demand for them. When customers and suppliers understand quality, the restraints and obstacles to achievement of quality excellence and Coascendancy are removed, beautiful products and services are produced, commerce is efficient, resources aren't wasted, employees attain pride in their ability to produce creative effort that

158

satisfies customers and, in time, quality of life for all continues its inexorable progression of improvement.

Strategy Switching

Creation of ascendant quality may be thought of as the creation of an attractive influence – where every customer seeking to perform the strategy addressed by the formulation of ascendant quality gravitates to the organization producing it. In this way, the producer of ascendant quality has the potential to attract most or all of the market for products or services serving this customer strategy. This form of attraction is a product of the phenomenon which the Process Enrichment philosophy defines as Strategy Switching.

STRATEGY SWITCHING

Strategy Switching is the process of changing strategy due to the recognition of the benefit of Ascendant Quality in an alternative strategy.

The realization that competition is a contest of strategies, leads to the significant insight that Strategy Switching is the primary cause for an organizations' loss or gain in customers. As customers exercise the capability to recognize what they value in a strategy, they can break free of the gravitational influence of their current strategy due to the attraction of the most valued strategic alternative. The term "Brand Switching" has served well in the past in describing this overall effect – but it hasn't served well in describing the cause of a loss or gain in customers. In contrast, the phrase Strategy Switching can lead to a more useful exploration of the causes of a customer loss or gain which must be found in one or more of the elements of strategy – including quality and risk.

A significant consequence of the availability of an improved strategy in a market is that enlightened customers will promptly recognize that, by comparison to the new standard of quality excellence, introduced through ascendant quality, all other strategies bear a risk of poor performance – since they are not as capable of serving the intended strategy. Organizations that continue to compete in this market will be compelled to either match or improve further upon the improved strategy and its elevated concept of quality excellence or else they will be pressured by customers' lack of demand to significantly reduce prices or compensate customers in some other way. In an efficient market, where better products and services get noticed, Ascendant Quality can serve to maximize an organization's profit as customers engage in Strategy Switching to meet their strategic interests.

Process Enrichment & Transformative Innovation

This is a true story. It's, in part, "a good action story" – but, as you'll see, it's mostly a drama. The tale begins in my college manufacturing technology class. The lab portion of this class was intended to give students hands-on experience with machine tools used in manufacturing. It was run much like a chemistry lab class where they give you a laundry list of things to do – and if you do it right… voila… you've created something. Just like in a chemistry lab, there was no real expectation that you'd deviate from the tried and true procedure.

I was watching one of the lab instructors as he set up a milling machine for our lab project – we actually got to cut metal in this class! I noticed that to adjust the milling machines' "headstock" – which is the upper part of this machine, he had to loosen a big hex nut that must've been awfully tight. To loosen it he needed a big wrench. Once the big wrench was on the big hex nut, he hit the end of the big wrench handle with a big hammer – because otherwise the big hex nut wouldn't budge. I started thinking about what I saw. I realized that thousands of machinists around the world were probably performing the same process the same way. I asked myself "Could there be another way to loosen a tight nut on a precision, state-of-the-art, machine without crudely hitting it with a big hammer?" Using my primordial concept of Process Enrichment – which has evolved to mean "working with the people in the system" – I almost immediately pictured an answer. If a pivot was made in the top of the wrench, once the wrench was firmly in place, it could be precisely hammered by swinging the handle into the top of the wrench. This would be better for the machine since it was a less reckless hit than a hit with a hammer and it was better for the worker because it was easier to manipulate one tool instead of two – and the placement of the tool itself helped make the hammer blows precise, controlled, and effective.

This one good idea I had as a teenager was appreciated by my father and we presented it to a patent attorney to see if it was patentable. The description of it I gave to the attorney became its name; "Wrench with pivotally mounted handle to induce hammer effect." Although my college professor didn't care much for my idea, it was a hit with many of the other students. Unfortunately for me – and also for my dad who paid for a patent search – what I thought was my invention had been invented, in 1921, by Oswald Duda. Apparently Mr. Duda has over a dozen patents for various types of wrenches. The sketch included in Mr. Duda's patent is shown in Figure 3-7. When I first saw Mr. Duda's sketch, I realized immediately that it was the same idea that I had – so, unfortunately for me, I didn't get to start my own industrial supply company. I would have loved to

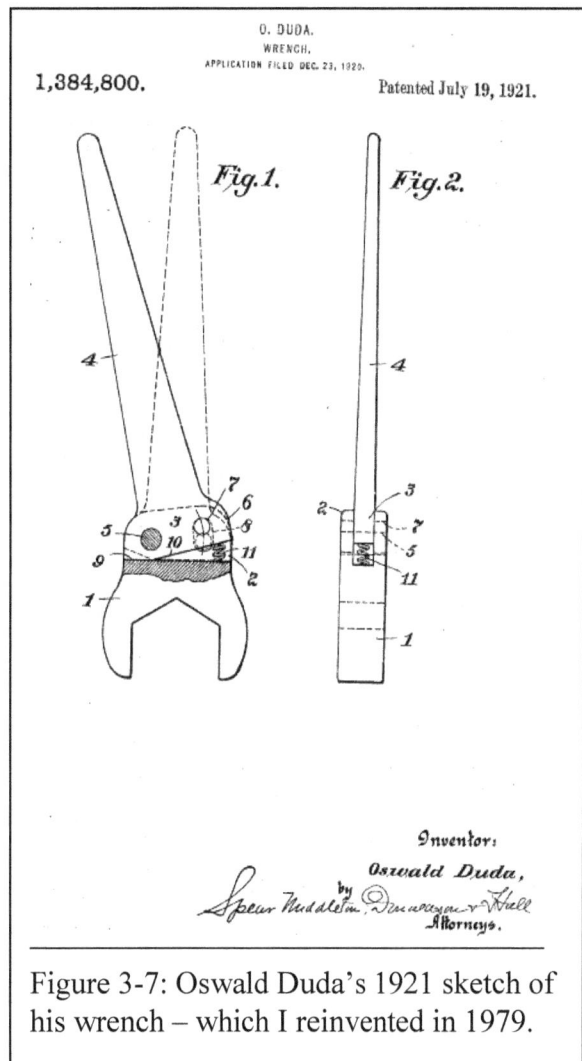

Figure 3-7: Oswald Duda's 1921 sketch of his wrench – which I reinvented in 1979.

find out how he came up with his (my) idea, if it was a success, and if his teachers or colleagues thought much of it.

In our effort to improve products and services beyond quality excellence and beyond the state of the art, we can begin by closely looking at quality excellence in each Theme of Performance and ask ourselves "If there was a position beyond the current level of quality in each Theme of Performance what would happen there? What would it take to get there? If we find that we've discovered a "Wrench with pivotally mounted handle to induce hammer effect." We may well have discovered transformative innovation that can serve to raise the standard of quality excellence to a new height.

We have now reached a plateau where the concepts of strategy, quality, and risk and their interrelationship is understandable and can be visualized. Our next challenge is to develop the essential wisdom with which this knowledge can be beneficially applied.

Case Study:

Operation Focus
The Air Battle at the Opening of the 1967 Six Day War

The challenge of creating effective competitive strategy and exceptional market value

* How do critical vulnerabilities influence innovation and the formulation of a competitive position that maximizes market value?

Introduction

The defining characteristic of competition is the offering, by competing suppliers, of creative effort bearing differing abilities to perform a strategy that a customer values. It is, thus, quality, as we have always known, which gives differing creative effort its differing market value. In consequence, a critical challenge for strategists, in the design of products and services, is to optimize the formulation of abilities to perform, in each of quality's six Themes of Performance, in order to offer customers desirable and recognizably superior performance over competing products and services. The extent to which the formulation of a competitor's products and services achieves quality excellence creates the relationship that establishes the creative efforts' competitive position. The Process Enrichment philosophy's definition, of competitive position, is presented here for the first time.

> ### Competitive Position
> *The competitive position of a suppliers' creative effort is a relationship established as the difference between the offered ability to perform a valued strategy and the ability to perform a valued strategy at the customers' standard of quality excellence.*

Our expression of competitive position as a relationship between actual performance and performance at the standard of quality excellence is useful as a means of communicating, in an influential way, specific details of how to improve performance, if necessary, so that a desirable competitive position can be attained. A means of visualizing this relationship is provided through the conceptual art of Saile Charting. On a Saile Chart, a competitive position is drawn as a battlefront – a zigzagging trench line through the Decisive Points in each Theme of Performance. Competitors' battlefronts can be compared to each other and to a battlefront representing the customers' ideal of quality excellence. Envisioning the quality of creative effort in the form of a battlefront, where a competition can be won or lost, has its virtue in vividly emphasizing the critical importance of the formulation of a competitive position to the achievement of competitive success.

When market offerings do not significantly differ in their competitive position and are roughly equivalent in their ability to perform a customer's strategy, a market is either a controlled market or it is a market in a form of stalemate, offering customers little inducement towards strategy switching. In commerce, this can be a desirable condition, for an extended period of time, generating stable revenue for competitors and acceptable products and services for customers who have not been attracted to innovative alternatives. In the case of military competition, however, where political visions are the basis of competition and where destructive behaviors are the requisite means to creating a condition with which those visions can be fulfilled, stalemate is an unsustainable, mutually destructive, condition which the adversaries should, logically, be motivated to resolve while their respective treasures still hold value. Under the circumstance of a destructive stalemate, almost any measures, even risky and extreme measures, which can energize the strategy of competitive warfare towards a decisive success, are considered welcomed innovations by people who understand and respect the value to civilization of ascendant quality.

Prologue

When adversaries employ significantly different strategic approaches to competition, this is a form of what the military calls "asymmetric warfare." Asymmetric warfare is reflected by the adversaries' significant differences in abilities to perform in one or more of the Themes of Performance. Among the most extreme manifestations of military asymmetric warfare is warfare between a nation and guerrilla fighters. In this scenario, the guerrilla force may use frequent small attacks on the center of gravity, or key strategies, of a nation in order to increase that nation's risk of poor performance. For example, in the revolt in

Israel against British Mandatory rule, the Irgun, an underground militia, identified and continually attacked Britain's center of gravity.

> "History and our observation persuaded us that if we could succeed in destroying the government's prestige in Eretz Israel, the removal of its rule would follow automatically. Thenceforward we gave no peace to this weak spot. Throughout all the years of our uprising, we hit at the British Government's prestige, deliberately, tirelessly, unceasingly. ...We believed in the supremacy of moral forces. It was our enemy who mocked at them. That is why, notwithstanding the enemy's tremendous preponderance in physical strength, he it was who was defeated, and not we. That is the law of history."[113]
>
> Menachem Begin
> Commander of the Irgun

The Irguns' policy of continual attack on British prestige – a treasured part of Britain's national identity and the "gravitas" which gave Britain the ability to influence world affairs – was implemented as a campaign of strategic communication accompanied by limited, but effective, warfare. The Irgun's policy of ridiculing British behavior was so effective on world opinion, in diminishing British prestige, that, according to Mr. Begin, "President Truman has said that he would have joined the 'terrorists' had he been in Eretz Israel during the rule of the British."[114],[115] Eventually, commando raids removing or destroying British weapons, accompanied by attacks on British prestige, served to compel the British to relinquish their mandate over Israel and, finally, in 1949, the refugees held in British concentration camps were freed. Asymmetric warfare was necessitated by the imbalance of resources and competencies between the adversaries. In the early 1940's it was possible, and also promising, for the Irgun to attack Britain's prestige, but the prospects for success weren't as promising, for the insufficiently armed Irgun, to form an army and openly battle the British –

---

[113] Begin, Menachem. *The Revolt* (Nash Publishing, 1978) p. 52.
[114] In The Revolt Mr. Begin explains that, historically, the Irgun were anti-terrorists and not "terrorists" because "Our purpose, in fact, was precisely the reverse of "terrorism." The whole essence of our struggle was the determination to free our people from its chief affliction—fear."
[115] *The Revolt*, p. 135.

who were better armed and experienced than ever before as World War Two raged on. As the example of the Irgun illustrates, under the right conditions, even a nation with a large, sustainable, military force can be defeated by a smaller, but an astute, audacious, and competent adversary when a nation bears the risk of critical vulnerabilities that can be exploited.

In battles governed by strategies created by adverse intellects, adversaries cannot always depend on overwhelming force to win. In a similar sense, successful, entrenched, companies should not believe that their market share is invulnerable to new competitors which may emerge and which may offer strategic performance that's more valuable to customers. Any organization can benefit from being attuned and responsive to the influence of ascendant quality to create movement first towards quality excellence in the lower echelon Themes of Performance, which should then drive progress through to quality excellence in the higher echelon Themes of Performance. In this way, improvement towards quality excellence and the drive towards Ascendant Quality beyond quality excellence minimize the possibility that risks of poor performance and critical vulnerabilities will be there to be exploited by competitors or adversaries. The product of this ascendant progression is a competitive position that we can recognize as market leadership.

In 1967, less than 20 years after the closure of holocaust-era concentration camps, Israel, at the time a nation of nearly 3 million people, a large portion of whom were survivors of that intensely tragic era, was completely surrounded by the combined armies of nearly a dozen Arab countries. Poised for attack, the Arab armed forces wielded approximately "900 combat aircraft, over 5,000 tanks, and a half million men."[116]

The charismatic leader orchestrating this overwhelming attack force was Egypt's President Gamal Abdel Nasser Hussein. Slick, impeccably dressed, in western style suit and tie, and adept in manipulating public sentiment, Mr. Nasser was a center of gravity of emergent Arab militancy. Unable, or unwilling to bring the progressiveness he, apparently, enjoyed personally to the people he led, Mr. Nasser sought instead, with the help of his Soviet and Arab allies, to be

---

[116] Oren, Michael B., *Six Days of War: June 1967 and the Making of the Modern Middle East*, 2002, p. 164.

the Arab leader who destroyed Israel. Armed by the Soviets, with state-of-the-art military equipment, Mr. Nasser pushed forward the idea of war and then pushed forward the acts of war. On the occasion of signing a mutual defense pact with Syria, Mr. Nasser expressed his vision in Egypt's nationalized press.

> "The armies of Egypt, Jordan, Syria and Lebanon are poised on the borders of Israel . . . . to face the challenge, while standing behind us are the armies of Iraq, Algeria, Kuwait, Sudan and the whole Arab nation. This act will astound the world. Today they will know that the Arabs are arranged for battle, the critical hour has arrived. We have reached the stage of serious action and not declarations."
>
> Gamal Abdel Nasser Hussein
> May 30, 1967

Syrian Foreign Minister Ibrahim Makhous, also fresh from signing the mutual defense pact with Egypt, spoke to the international press. Mr. Mikhous explained the purpose of Syria's alliance with Egypt and the meaning of Mr. Nasser's demand for the UN forces keeping peace along the borders of Israel to leave.

> "Our two brotherly countries have turned into one mobilized force. The withdrawal of the UN forces ... means 'make way, our forces are on their way to battle."
>
> Ibrahim Makhous
> May 30, 1967

If being completely surrounded and outnumbered by heavily armed adversaries wasn't bad enough, the leaders of neutral or friendly countries demanded inaction from Israel – even as Israel continued to be surrounded – as a pre-condition for any potential support they may provide. For example, according to the U.S. State Department, "Though [U.S. president Lyndon] Johnson continued to caution Israel against preemption, a number of the President's advisors had concluded that U.S. interests would be best served by Israel "going it alone" by the time the Israelis actually did so."[117]

---

[117] U.S. Dept. of State, http://history.state.gov/milestones/1961-1968/arab-israeli-war-1967

Although President Johnson expressed that Egypt's blockade of the Gulf of Akaba[118] to Israeli shipping "is illegal and potentially disastrous to the cause of peace,"[119] and, although he deplored the build-up of military forces, the president's attitude remained indifferent – even after full scale war broke out. On the first morning of the war, at a June 5th, 1967 White House press conference, Dept. of State spokesman Robert J. McCloskey said to reporters "Our position is neutral in thought, word, and deed." A reporter then asked, "Do you feel we can continue to maintain a neutral position, no matter what happens in the Middle East?" and Mr. McCloskey replied, "That will be our effort."[120]

Neutrality, from the U.S. viewpoint meant, among other things, that following the outbreak of war, the U.S. would not provide Israel with military equipment or supplies. France and Britain also joined the U.S. in a complete arms embargo against Israel. Meanwhile, the Soviet Union, which, only 19 years earlier, voted in the United Nations in favor of creation of the state of Israel, continued to supply and replenish the armaments of the Arab armed forces in the vain hope that if Israel was destroyed the Arab nations would be grateful.

As the advancing Arab armies took up fortified positions along Israel's border, there was no doubt that war was imminent, and, apparently, Mr. Nasser felt comfortably in control of when the Arab attack would begin in full scale. His adversary was surrounded, isolated, and equipped with a limited supply of foreign equipment, which the embargo ensured they couldn't replace, while his own forces, present in overwhelming numbers, enjoyed the active materiel support of a superpower. As Mr. Nasser could assure himself, quality excellence in his war strategy was achieved by conformance to the requirements of textbook military doctrine for victory in warfare. With this achievement, Mr. Nasser must have enjoyed a great sense of fearlessness and the conviction that a glorious victory was within his reach.

The secret and brilliant military plan that triumphed over Mr. Nasser's plans for destruction and glory was code named "Operation Focus."

---

[118] "In 1957, President Dwight D. Eisenhower had promised that the United States would treat the closure of the Straits as an act of war." U.S. Dept. of State, http://history.state.gov/milestones/1961-1968/arab-israeli-war-1967
[119] Johnson, Lyndon B., U.S. President. May 23, 1967
[120] Califano, Joseph. *Memorandum from Joseph Califano to the President*, June 5; Johnson Library, Appointment File, Middle East Crisis, June 1967.

Operation Focus: The Air Battle at the Opening of the 1967 Six Day War

Operation Focus was a product of Israel's strategic planning that properly anticipated the gravity of the onslaught that the combined Arab militaries could muster as well as the limited resources the Israeli Defense Forces would have available. The inability to sustain a protracted war, against a larger attacking force, was the intractable critical vulnerability which the Israeli Defense Force's strategy was compelled to compensate for. In order to protect its cities from enemy bombers, the vital circumstance of any war must be for Israel's Air Force to maintain air superiority over Israel. Israel recognized that if the air war was conducted as a series of dogfights, there could be a significant risk of attrition which might give an advantage to their Arab adversaries who started the war with a 3 to 1 advantage in military aircraft and who also enjoyed a ready capability to be resupplied by their Soviet allies. In light of the failure of diplomacy, what Israel needed was a different strategy which would, somehow, keep their adversary's planes on the ground where they could remain harmless. The intention to contend with these exigencies was at the genesis of Operation Focus.

In Operation Focus, general Mordechai Hod, commander of Israel's Air Force, in June of 1967, brought to perfection an audacious plan, originally developed by his predecessor, general Dan Tolkovsky, in the 1950's, requiring the simultaneous use of all of Israel's military aircraft, for a devastating first strike, to destroy all 18 of Egypt's forward airbases. Once the runways at each airbase were destroyed, enemy aircraft could not take off or land and were vulnerable to attack on the ground. In this way, air superiority over Israel, as well as over the battlefields, could be secured. The strategic intention to destroy enemy aircraft on the ground has been called the Hod doctrine. In general Hod's words, "If you have to resort to dogfights you have failed, for you should get them on the ground."[121]

Details of Operation Focus were kept mostly secret, even from the politicians who authorized its implementation; however, strategic planning and intelligence gathering as well as training for Operation Focus were relentlessly optimized over the years before the outbreak of war. In this way, Operation

---

[121] Rodwell, Robert R. "Three Hours and Six Days," Air Force Magazine. October 1967, 59. Quoted in Ltc. Ronald D. Jones, Israeli Air Superiority in the 1967 Arab-Israeli War: An Analysis of Operational Art, Naval War College, Newport, R.I.

Focus, was designed to compensate for the Israeli Defense Force's critical vulnerability in the Sustainment Theme of Performance by achieving and ascending beyond historical precedents of quality excellence in doctrine by introducing innovation to a high degree in the higher echelon Themes of Performance.

In the Optimization Theme

- A capability was developed to simultaneously fly the entire air force, just above the trees on land and just above the waves at sea, in order to evade detection by radar, and, also, to fly in precisely maintained groupings and intervals to reach and destroy at least 11 targeted airfields simultaneously in the first wave.

- A historian described the "Concrete Dibber" as "an ingenious bomb which the Israelis have devised and perfected for the specific purpose of destroying runways. As soon as the bomb leaves the aircraft a retrorocket is fired to stop its forward impetus. Then a booster rocket drives it into the runway. Once it has penetrated the concrete a time fuse explodes it. The fuse may be instantaneous or may be set on a variable time-delay. Normally runways are considered easy to repair, but it is rather more difficult when the runway keeps exploding. The point of this unique bomb is to enable an aircraft to bomb runways while flying at low level and high speed. A conventional bomb released in this way would bounce and do only superficial damage."[122] The ability to drop the new bomb at high speed and at low altitude allowed Israeli aircraft to stay below the altitude where enemy antiaircraft fire could be effective.

- Time on target, for all 11 airbases struck in the first wave, was coordinated to be 8:45 AM because "around 0800 the weather is usually at its optimum. The visibility is at its best because of the angle of the sun, and the air is at its stillest, which is important when it comes to placing bombs accurately on runways."[123]

- In order to conserve fuel, to reach distant targets, some twin engine aircraft had to fly much of the way using only one engine.

---

[122] Churchill, Randolph S., and Winston S. Churchill. *The Six Day War*. 1967. Reprint. London: William Heinemann Ltd., 1967, p.84.
[123] *The Six Day War*, p.79.

In the Process Enrichment Theme

- Intelligence on the dispositions of enemy forces and the habits of pilots was gathered so that the mission occurred at the ideal moment of target vulnerability.
- The Israeli air force developed a process and trained crews so that they were capable of rearming and refueling a fighter jet returning from combat in under 8 minutes from touch down to take off. "Compared with the ground turn-around time of 7-10 minutes of the Israeli Air Force, that of the Egyptians was more than two hours."[124]
- "Public opinion in most western countries had come round to the side of the Israelis when it saw the recklessness and provocation of Nasser's action coupled with Israel's reluctance to take decisive action without first seeing what diplomacy could achieve."[125]

In the World-class Theme

- Development of a strategy to destroy nearly the entire air forces of three nations, in close succession, while, also, providing air support for one of the most intense tank battles in history which occurred in the blistering heat of the Sinai Dessert.

The effect of these high Theme innovations was to create a force multiplying destructive capability never before or since executed with such devastating results. The opening phase of Operation Focus was described by historians.

> "As the first wave of Israeli aircraft struck their targets, the second wave was already on its way and the third wave had just got airborne. They were spaced at 10-minute intervals. Each flight of four aircraft were given seven minutes over their targets – enough for three or four passes, one bombing run and two or three strafing passes. An extra three minutes was allowed for navigational error or for an extra run over the target. The Israelis were operating on an incredibly fast turn around time. For aircraft striking the main

---

[124] *The Six Day War*, p.89.
[125] *The Six Day War*, p.70.

Egyptian bases in the vicinity of the [Suez] Canal the rotation would have been as follows:

- Time to target: approx. 22 ½ minutes.
- Time spent over target: approx. 7 ½ minutes.
- Return to base: approx. 20 minutes.
- Ground turn around time: approx. 7 ½ minutes.
- Total: approx. 57 ½ minutes.

This means that the Israeli aircraft were over their targets for the second time within an hour of their first attack."[126]

Randolph S. and Winston S. Churchill
The Six Day War

As with most innovative ideas, and especially in an environment where adverse intellects challenge each other, there is always uncertainty about how the risks of a strategy will play out in the real world. This sentiment was expressed by General Hod in an interview with the Jerusalem Post in which he said "Everything depended on our fooling the Egyptians for the 45 minutes until our first planes reached their targets... These were the longest 45 minutes of my life."[127] After these 45 minutes, as the first wave returned and the report of the destruction of over 300 first-line enemy fighters and bombers was revealed, the ultimate outcome of the war could not be in doubt. Due to Operation Focus, after only 45 minutes of full scale war, Egypt's defeat was virtually inevitable.

The competitive positions of the adversaries are depicted in the Saile Chart of Figure 3-8. Since the equipment of both adversaries was nearly state of the art, high ratings for performance in the Acceptance and Efficacy Themes are appropriate. The first big difference in competitive position is in the Sustainment Theme. With the enormous 3 to 1 numerical advantage in aircraft the Arab air forces enjoyed, they could afford to take some risks and endure some losses. On the Israeli side, the imbalance in the number of aircraft made it essential that they compensate for this risk in the Themes of Performance where they could. The most pronounced compensation was in the drive to make sure the war was

---

[126] *The Six Day War*, p.82
[127] Rabinovich, Abraham. The War Nobody Wanted, Jerusalem Post, 6/13/1997, retrieved 5/20/2014, http://archive.today/7U4Jd

won quickly – before risk in the Sustainment Theme turned to their enemy's advantage. The goal of winning quickly, which originated due to the enlightening influence of the cognizance of risk, required competencies well beyond quality excellence according to traditional doctrine. In innovation, and in transformative innovation, which shaped Israeli strategy, leaders played their role actively. As General Ezer Weizman, Director of Operations said "... plans are like bricks. They can be used one by one to build up a structure as the situation develops. We don't go in for preconceived and, therefore, inflexible master plans."[128]

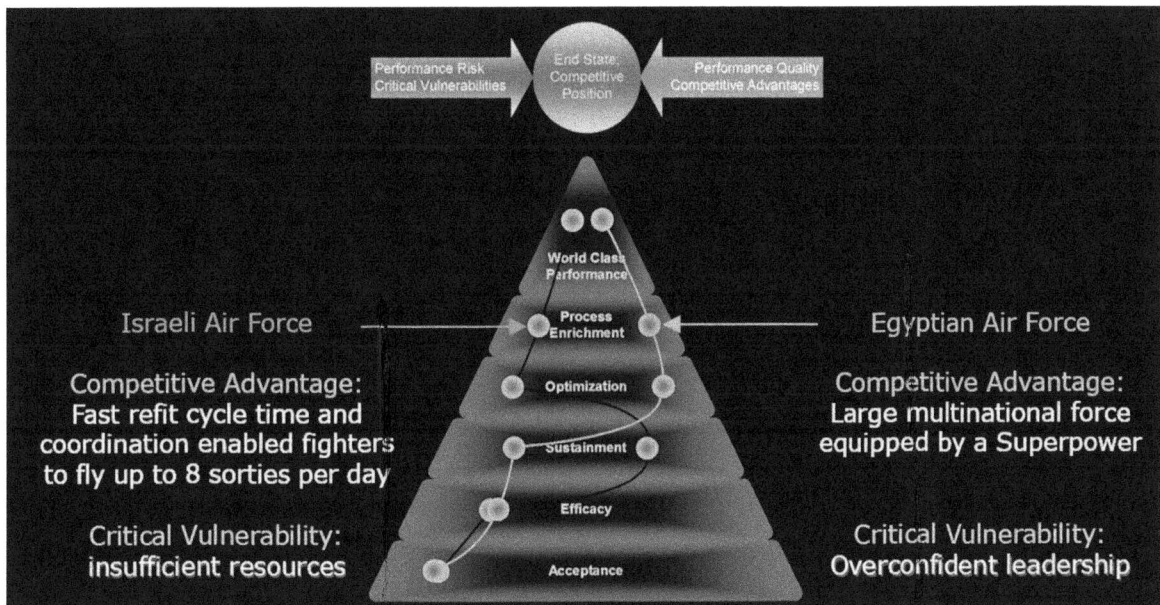

Figure 3-8: An evaluation of adversaries' competitive positions at the beginning of the 1967 Six Day War.

After the first wave of Operation Focus, which devastated Egypt's air force, President Nasser, for reasons only he can explain, informed his Arab allies that 70% of Israel's air force had been destroyed. On this basis, at around noon on June 5th, Syria and Jordan joined Egypt in the war by sending their air forces to attack targets in Israel. Since, by this time, the Egyptian air force was largely incapacitated, planes that Israeli ground controllers planned to send against Egypt were, instead, diverted to attack the airbases of Syria and Jordan as they had done in Egypt. Referring to the diversion from a targeted airfield in Fayid, Egypt, to an airfield in Amman, Jordan, one historian observed, "The ability to

---

[128] *The Six Day War*, p.65.

divert aircraft to a new target, which was over 230 miles from the original target, reflects a high degree of operational flexibility."[129] Operational flexibility is an aspect of performance that we associate most closely with the World-class Theme of Performance. There can be little doubt that the cumulative effect of quality excellence in every Theme of Performance except the Sustainment Theme served to create quality excellence in the World-class Theme. The impact on the war of the Israeli air force's World-class performance was expressed by General Hod "We were able to deal with Syria and Jordan in twenty-five minutes."[130] As depicted in Figure 3-9, the ability to replicate the destruction of airfields in Egypt, Syria, and Jordan, in Iraq, on the second day of the war, in response to the Iraqi air force's attack on a civilian Israeli town, reveals the inestimable market value of World Class performance.

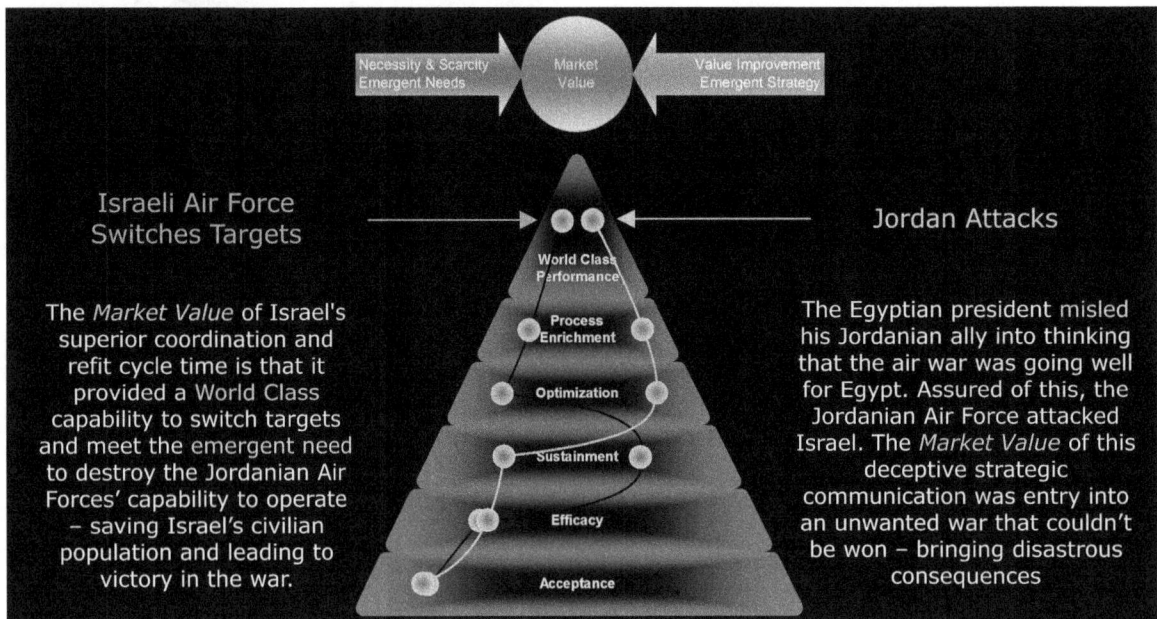

Figure 3-9: An evaluation of adversaries' capabilities in the World-class Theme of Performance during the 1967 Six Day War.

---

[129] Douglas, Charles. *The Arab-Israeli War - II* (Royal United Institution Journal, Vol. 112, No. 648, November, 1967), p. 335. As quoted in: Maj. Stanley S. Gunnerson, A study of Airpower Employment in the Six-day War, Air Command and Staff College, May 1971.
[130] *The Six Day War*, p.86

"In the first two days of the war the Israeli Air Force flew more than 1,000 sorties, many of the pilots flying up to 8 sorties a day."[131] "Up to nightfall on the second day of the war, the Israelis had destroyed 416 aircraft, 393 on the ground."[132] The Hod doctrine had been followed well. An equally incredible performance of the Israeli Army, said by historians to be "one of the finest armies the world has ever seen"[133] compelled all the adversaries to sign a cease fire agreement under the auspices of the United Nations.

In the end, President Nasser, confronted with an inglorious defeat, acknowledged the effectiveness of the Israeli Air Force in his resignation speech (he withdrew his resignation the next day).

"It can be said without fear of exaggeration that the enemy was operating an air force three times its normal strength."[134]

Gamal Abdel Nasser Hussein
June 9, 1967

Epilogue

In the end, Operation Focus was undoubtedly a spectacularly successful strategy of air war. In Operation Focus, the ratio of enemy planes destroyed was approximately 23.7 to 1 – an astonishing and unprecedented military performance and an enlightening example of the application of innovation and ascendant quality to competitive strategy – in light of a significant and intractable critical vulnerability in the Sustainment Theme of Performance.

As in the Battle of Britain, the secret to Israel's success was in the number of sorties, and also primarily in the audacity and meticulous planning and execution of their strategy. Leadership also served to energize the strategic competencies of the Israeli Air Force even as the command changed from Dan Tolkovsky to Ezer Weizman and then to Mordechai Hod. The enduring strategic motive that served to challenge the Israeli air Force to perfect their strategy of war was expressed by General Weizmann: "We always went on the assumption that

---

[131] *The Six Day War*, p.88
[132] *The Six Day War*, p.87
[133] *The Six Day War*, p.60
[134] *The Six Day War*, p.89

we would be fighting the finest air force in the world – then we set out to show that this was not the case."[135]

The reasons behind the 1967 Six Day War, or for any of the preceding or subsequent wars between Israelis and Arabs, cannot be entirely explained logically, although, if peace is to be attained, it may be attained through the entirely logical and very practical advent of the realization that the Israelis and Arabs of the Middle East would make great, progressive, allies. In many cultural developments and business ventures, where Arab states are rich in resources, but weak in technical competencies, Israel has it the other way around. This asymmetry in peace is also at the root of the adversaries' asymmetry in war. This truth has evaded the platform of political leaders in the Middle East since the days of Emir Faisal, the King of Syria, who wrote of his effort, along with Chaim Weizmann, leader of the Zionist movement, "to encourage and stimulate immigration... into Palestine [part of Syria in 1919] on a large scale."[136]

The Coascendant purpose of encouraging immigration is evident in a letter Emir Faisal wrote to Felix Frankfurter, an American Zionist, during the Paris Peace Conference following World War One.

"We are working together for a reformed and revived Near East, and our two movements complete one another... and there is room in Syria for us both. Indeed I think that neither can be a real success without the other."[137]

Faisal
March 5, 1919

The presence of the basis for Coascendancy is everywhere to be discovered for everyone who chooses to resolve unnecessary conflict in pursuit of a greater purpose.

---

[135] *The Six Day War*, p.66
[136] Faisal-Weizmann Agreement, 1/3/1919
[137] Weizmann, Chaim. The Letters and Papers of Chaim Weizmann: August 1898-July 1931, p. 237

Chapter 4:

The Nature of Risk

From their earliest days of emergence into life
to their conquest of centuries of survival
even the most humble of trees
will continuously fight
any and all risks
to which they are exposed –
in any way they can –
to ensure a life of quality excellence.

Photo: "*Home, at last*" provided courtesy of Elizabeth Norman.

Chapter 4:

The Nature of Risk

PERSPECTIVES

- ▶ The Wisdom of a Tree

- ▶ The Nature of Risk

- ▶ The Lifecycle of Risk

- ▶ The Process Enrichment philosophy's definition of risk

- ▶ Exercise:
  Composing a Risk Management Strategy to Circumvent and
  Counteract the Risk of Improvised Explosive Devices Emplaced
  in Roadways

- ▶ The Fearlessness Effect

  - ▪ The Parable of the Fearless Slugger
    An exercise in the flaws, fallacies, and limitations of
    contemporary methods of risk assessment

  - ▪ Case Study:
    NASA's Space Shuttle Program and the Fearlessness Effect

    How the influence of a governing strategy interacted with a
    flawed understanding of the conceptual system of strategy to
    cause both of NASA's Space Shuttle accidents.

- ▶ Thematic Risk Exposure Evaluation (TREE)

- ▶ How the enlightening influence of risk can be used to drive
  improvement of quality, strategy, and competitive position

In our quest for ascendant quality we strive to achieve, sustain, and grow our ideal of quality excellence progressively because we can readily appreciate the beauty in quality excellence and in the fruits of innovation which improve our quality of life. While the rewards of quality improvement entice us and challenge us to contemplate the prospects for improvement within our ability to imagine them, risk, which often propels adversity in its wake, bears the sobering influence that can be decisive as to whether or not an innovative idea can be advantageously implemented as part of a successful strategy. Although the prospect of risks' adverse effects can inhibit our progressive inclinations, an ability to assess the elements of risk, sufficient to develop strategy to circumvent the risks' adverse effects, and thereby overcome their restraining influence, can serve as a talent that improves our potential for safe, innovative, improvement. Like the professional Grand Prix driver who has walked through every hairpin turn and every gradient change of the race course and who can then foresee the highest speed the road will allow, and where on the road the highest speed is possible, when we deepen our knowledge of the risk in our creative endeavors we can foresee how and where we may adapt our strategies to improve performance and confidently drive innovative progress faster. The ability to adequately assess risk's elements, and their influence on strategy, is an indispensable talent that enables us to gain the motivation to achieve the quality excellence and progress in our creative effort which we truly desire while it also serves to reduce or avoid the potential for the adversity of risk in everything we do.

The nature of risk and the way various risks we encounter influence our lives can be understood with the guidance of the wisdom of a tree. A tree that has grown under conditions of quality excellence is, undoubtedly, a beautiful tree because the healthy development and the achievement of beauty in a tree are largely a product of the environmental conditions to which a tree is exposed. We can conclude that most, if not all, of the risks to a tree come in the form of adverse environmental conditions which put the tree's ability to attain the nourishing resources it needs in jeopardy. Risks to a tree that are present due to environmental conditions are risks that a tree must respond to with great wisdom, because these are not only risks to a trees' quality of life, these are risks to a trees' ability to survive, and these are, also, risks which a tree cannot avoid. The wisdom of a tree is evident in a tree's behavior, under conditions of risk, when, for example, a tree living in a dry climate broadens and deepens its roots until quality excellence in the capability to absorb the water and other nutrients it needs is achieved. The wisdom of a tree is also evident in a tree's response to the risk of insufficient sunlight. In this case, quality excellence is restored when the tree exercises a strategy of aligning its leaves with the sun in order to absorb enough sunlight to produce the energy it needs through photosynthesis. In the behavior of a tree, it is evident that a tree's quest for ascendant quality is supported and advanced by the formulation and exercise of strategy that acts to circumvent, counteract, and triumph over risk by restoring, sustaining, and maximizing quality in response to adversity in the environmental conditions to which the tree is exposed. The wisdom of a tree tells us that the nature of risk is that it can be an enlightening influence which can serve to improve our strategies if we understand and respond to its ramifications for our survival and success.

The sophisticated nature of the conditional risks which a tree experiences, are very different from the relatively simple, predictable, risks which we are often inclined to bring upon ourselves. In our creative effort, for example, we often speculate on risks that have the same basic structure and simplicity of a bet on a coin toss. In a coin toss we know that we have a 50% chance of heads or tails which determines whether we win or lose a specific incentive. In speculation, of this form, which we cause by our pursuit of a financial incentive, we can easily limit our potential financial losses and we can decide whether or not we wish to chance the risk. In the wisdom of a tree, however, there is no inclination to speculate on the complex and diverse conditional risks that may threaten their survival. We, on the other hand, are willfully, and often excessively, inclined towards speculative risk. In the interest of attaining incentives through our creative effort, we find in risk an opportunity to less expensively, more easily, or more quickly, attain desired incentives and we are, therefore, willing to consider putting at risk just about anything we value, under many different conditions. We speculate on innovative and efficient production technology to meet market demand, reduce costs, and increase profit; we speculate on advertising to create improved market performance and profit; and we also speculate on scientific discovery to improve our potential or in the service of causes with the potential to benefit humankind. When we are truly inspired to attain a valued incentive, particularly an incentive perceived to have a Coascendant purpose of advancing civilization in some way, our need to be on the quest for ascendant quality will drive us to put our valued treasure and even our lives or the lives of others on the line, in order to attain our goal. These forms of speculative risk taking are, of course, things a tree would never do. Our wisdom with respect to incentive-driven risk taking is often in conflict with our natural inclinations, and natural wisdom – yet it is with our distinctly human form of speculative risk taking, experimentation, and our quest for ascendant quality that our creative effort reflects our drive to reach beyond the trees and beyond the stratosphere and into unexplored realms, while the wise tree enjoys a contented life that, for some trees, can span several thousand years.

The wisdom of a tree and our quest for ascendant quality, which have provided us insight to the causes and nature of risk, and to risk's influence on our thoughts and behavior, are the basis for a new definition of risk provided by the Process Enrichment philosophy and introduced here for the first time.

<u>RISK</u>

Risk is an influence affecting strategy caused by an incentive or condition that inhibits transformation to quality excellence.

When a strategy deviates from quality excellence, risk is continuously present as an enlightening and beneficial corrective influence serving to drive improvement of conditions or serving as a cautionary influence to prevent a strategy of excessive speculation. The idea that "risk is an influence" may, at first, be difficult to grasp, however, evidence that this is the case can be recognized by considering the circumstances which our culture readily identifies as risk, such as:

- In an old cartoon, a villain tied Nell to train tracks and she was *influenced* to struggle and to cry out for help as she expected a train to arrive. Since she was tied up, Nell couldn't transform her condition to quality excellence – meaning safety in this case – and so she remained at risk. Dudley was *influenced* to rush to the scene in time to untie the ropes and save Nell.

- At the baseball game a fly ball was high in the air and looked like it was about to come down and hit a fan, but the fan was *influenced* to move out of the way. Another fan, who brought a baseball glove and valued a game ball, was *influenced* to accept the risk, and moved into the path of the ball and caught a great souvenir.

- If one of the engines on an airplane fails during flight, then the airplane may crash, and so the pilot is *influenced* to adopt a counteractive strategy to balance the plane and improve the quality of the condition of flight. The pilot will also be *influenced* to land the plane at the nearest airport.

- If we are running out of money while on a strategy of vacation, we may be *influenced* to spend less money or return to a strategy of work.

- If we run out of money to buy food then we will go hungry and so we are *influenced* to alter our strategy and earn or somehow acquire money to improve our condition.

- If a child is "at risk," due to conditions in the home environment, then relatives or social workers may be *influenced* to intervene to develop strategy to remove the barriers to the child's proper development.

- If a gambler takes chances at the roulette wheel with his month's rent then the *influence* to attain the incentive of the potential winnings is stronger than the *influence* of losing the rent money and perhaps not being able to live indoors.

Since the effect of risk is to compel a reasonable person to think of, and often act on, improved or alternative strategies, in order to ensure that the strategy pursued is nearer to quality excellence, we can confirm that risk is an influence affecting strategy.

In the use of the word "influence," to define risk, we find a rather telling link to the metaphor of strategy acting like a center of gravity. The gravitational energy which influenced decision makers to be attracted to a strategy in the first place is virtually identical to the energy, or influence, created by the recognition of risk and the desire to improve because both influences are endowed with the virtue of serving the same ascendant purpose of the same strategy. In recognizing risk we gain the enlightening wisdom and corrective influence necessary to restore, improve and thereby energize a strategy so that the ascendant purposes of the strategy can be pursued with greater confidence in success.

As our definition of risk indicates, the cause of any risk is that there is either an incentive in a strategy which makes risk taking desirable or there is a condition, that hasn't been overcome, which introduced risk to the strategy or there are a combination of both incentivized and conditional causes of risk. The insight that, within a strategy, there are only two causes of any risk is helpful in focusing the effort to troubleshoot any risks that we may encounter. We may also improve the precision of our understanding of any risk, and the risks impact on the overall strategy, by realizing that any risk will be found to affect performance within one or more of quality's six Themes of Performance. The process of thinking through how an organization, a project, a process, a product or a service performs or is designed to perform, in each Theme of Performance, should expose any deviations from the standard of quality excellence or from the desired standard of performance. These deviations can be recognized as risks and should be tracked, through progressive improvements, until performance achieves the intended standard of quality. A subtle, yet significant, philosophical point is that when we seek to remediate, mitigate, remove, or resolve risk, what we are actually seeking is to implement an initiative, in the spirit of the quest for ascendant quality, to improve the strategy governing the risk in order to approach or attain quality excellence.

On the occasion of defining the concept of risk, it is important to reflect on the significance of this momentous turn of the third and last tumbler of the combination lock which we needed to open to reveal the full conceptual system of strategy, quality, and risk as they are viewed from the perspective of the Process Enrichment® philosophy. In our new conceptual system, in strategy we gain our incentives, motives, and our intended course of action; in quality we find the abilities to perform a strategy with which our creative effort becomes desirable to customers; and in risk we find the enlightening influence to make strategic course corrections as we pursue the quest for ascendant quality. This conceptual system when paired with the compatible technology of Saile charts, introduces a capability to visually evaluate and communicate useful information about the impact of risk on the quality of an entire system and on the strategy for using that system which was not possible before.

To appreciate the differences between the traditional viewpoint on risk and risk from the perspective of the Process Enrichment philosophy, I hope you will enjoy the Parable of the Fearless Slugger.

The Parable of the Fearless Slugger

A famous baseball slugger, unfortunately, was hit in the head by a wild pitch during an important playoff game. He was knocked to the ground, but he recovered immediately from the physical injury and jogged to first base to the cheers of the crowd. On his next at bat, however, the slugger was afraid of every pitch and he struck out rather pathetically. The team manager immediately realized that unless the slugger quickly overcomes his fear, the playoff game will be lost and the team will be out of the World Series. The radio commentator broadcasting that day and many of the fans also realized the psychological nature of the problem. A relatively well dressed fan, watching the game from the grandstand, approached the team manager. He said, in a European accent and with a heavy stutter, "I c-can c-c-cure the s-s-slugger. I'm a psych-ch-ch..." and before he could finish the sentence the team manager said "Great Doc! Here he is. See what you can d-d-do." The manager wasn't poking fun – it's just that the weather was rapidly getting cold. It was already about 40 degrees and it was expected to go down to freezing in an hour or so due to a cold-front moving in.

"Doc," as he's now called, say's "f-f-follow me" and he leads the slugger and the team manager to the players' practice area where there's a batting cage. The batting cage has four walls and a roof all made out of a wire fence so it's easy to see inside. Everyone gathers around, as more and more curious folks start to appear. Doc places a chair on home plate, in the batter's cage, facing an automatic pitching machine. A maintenance engineer for the ball field happens upon this group and the Doc asks him "C-c-can you s-s-set up the automated pitching machine to p-p-pitch exactly to the s-s-slugger's f-f-forehead as he sits in the ch-ch-chair on home p-p-plate?" The engineer exclaimed, "That's c-c-crazy! The pitching machine throws mostly slow balls – which can hurt badly enough – but there are, also, 2 fast balls mixed in with every ten pitches – and they will definitely kill the slugger if he gets hit by them!" The Doc replies, "That's why we're here, f-f-for Fearlessness Training – every d-dangerous pitch the S-Slugger survives will improve his f-f-fearlessness and he will be back in the game after only

3 pitches. Besides, he will be wearing two helmets. If the first helmet cracks, the second helmet will p-protect the s-slugger. Also, he will b-b-bob his head from s-s-side to s-side to avoid the ball. C-C-Capish? Understand?" The team manager chimes in, and introduces himself, saying "I'm the team manager..." and he demands of the engineer "Does this set-up look safe to you?" Thinking quickly, the engineer says "I don't know how well the slugger can bob his head, but I think two helmets will be very safe if they stay on – but I'm too cold to argue with you. It's your decision."

As the slugger gives the team manager a wink, Doc nods an ok, and the engineer looks on for the team manager's decision, the team manager says "Let's go! We've got a game to win!" And so, into the batting cage goes the brave slugger. He sits there facing the automatic pitching machine as Doc secures the two helmets on him. When Doc says "Ready boys" the team manager nods and flips the switch turning the pitching machine on. The gears on the machine began their purposeful churning around which snapped out the pitch. The first pitch was a slow ball and the slugger bobbed to the side safely – but the helmets fell off. The slugger immediately jumped to his feet and exclaimed: "This is great – I feel better already!" The small crowd – which was growing larger – roared and cheered! The Doc, smiling broadly, quickly fixed the two helmets back on the slugger.

The engineer agitatedly interrupted this process to say, "If the helmets fall off before the pitch, and if the pitch is a fastball, the slugger will get killed!" The manager, challenging this alarm, asked "Are you saying the slugger will definitely get killed on the next pitch?" The engineer replied "No, there's no certainty of a catastrophe – he certainly survived one pitch already – but there's certainly a slight risk of a catastrophe. Here is my standard risk matrix for the next pitch..."

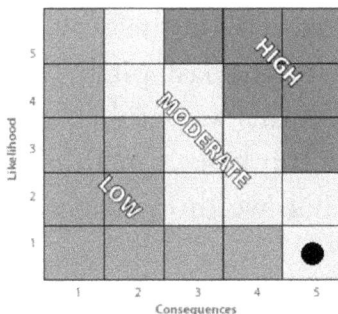

Figure 4-1: Initial risk assessment for Fearlessness Training using a traditional risk matrix.

The manager said "OK, I see there's a consequence of 5 which is as high as you can get – for loss of life – but the likelihood is a 1 which is the lowest you can get. This risk is in the "moderate" risk category." The manager asks Doc and the engineer "Can you accept this risk?" and they both reply in the affirmative. The slugger isn't told about the risk because it's only a moderate risk. The manager says, "I accept this risk too. Let's get on with it!" And, for the second time, the manager flips the switch turning on the automatic pitching machine.

The slugger takes the second pitch, which is, again, a slow ball – but this time it hits him, not quite directly, but a serious glancing blow to the side of the top helmet which makes both helmets fly off together into the fence. The top helmet isn't broken, but it has deep scratches and a dent that cracked through just a bit. The second helmet is also deeply dented and scratched up a little. The slugger, once again, jumped to his feet and shouted – "It's working! I feel great!" and the ever increasing crowd roared and cheered. Then, getting the slugger's attention, the Doc asked "Why didn't you m-move out-out of the way f-faster?" and the Slugger answered "Well, Doc, I was sh-shivering a little from the cold. But boy do I feel great now!" The temperature was now just above freezing. The Doc grabbed 2 new helmets and, again, fixed them on the slugger. The engineer put his hand on the Doc's shoulder and said "You can't let this go on; both helmets were damaged, in ways they weren't designed to be damaged, and both helmets fell off at the same time. There's not much redundancy in the safety precaution – and the fast ball will certainly kill the slugger! Capish? Feshtay? Understand?" So the Doc say's "Look how happy the slugger is and how happy the crowd is and he's already getting fearless with two pitches and he only has one pitch to go." While Doc and the engineer are arguing, the manager steps in and asks the engineer "Can you quantify your concerns?" And the engineer thinks to himself for a moment and says "I can't. I have no data to quantify it." And then the engineer says "Because it has gotten much colder now, and the slugger got hit because he shivered, instead of bobbing out of the way, the probability of success is away from goodness compared to the last pitch." The team manager says "We know the Slugger is a top athlete – this was probably just a one-of-a-kind event." The engineer responded "I don't think so. I'm thinking the probability of a catastrophic consequence is higher than we thought because of the cold. Here's the updated risk matrix."

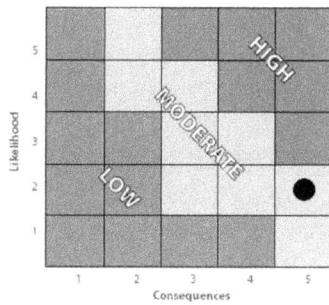

Figure 4-2: Updated risk assessment for Fearlessness Training using a traditional risk matrix.

Looking at the updated risk matrix, the team manager said to the engineer "I see you've increased the probability of occurrence of the risks' consequence to 2 which is the second lowest rating and the consequence of the risk is still 5. According to your risk matrix the risk is still only a moderate risk – just like last time. We've had 2 successes that nearly cured the slugger – one more pitch will restore all his fearlessness and will win us the playoff game and get us into the World Series." The team manager continued... "The Doc put helmets on the slugger that are good as new – now that's quality." and then he asked the engineer "Can you prove the next pitch will kill the slugger?" and, after a searching pause, the engineer answered "No. But..." Cutting off the engineer in mid thought, the manager said to the engineer "Take off your engineer's cap and put on my manager's cap and think like a manager about all the people counting on the slugger and then tell me if you accept the risk." And so, the engineer accepted the risk – so did doc and so did the team manager. The slugger wasn't told about the risk because it was, still, only a moderate risk. The team manager asked the slugger "Are you ready for the next pitch?" and the slugger said "You kn-n-now it s-s-sure is c-c-cold out here. But if you're... you're ready to go so am I!" Once again, a roar of cheers went up from the crowd as the slugger settled into his seat for the next pitch and Doc again fixed the helmets on him carefully.

On the doc's ready signal, the team manager switched on the automated pitching machine. A hush came over the crowd and the gears of the big machine could be heard grinding around, tightly coiling the powerful and precise springs, working up to a tense silence and, suddenly, an explosive roar and then – out came the pitch. The third, and last, pitch of the slugger's life was a fast ball, which, on that freezing evening, he faced bravely, and fearlessly.

Questions for Consideration

- In your opinion, was the risk acceptable before each of the three pitches?

  o Did you reach your opinion quantitatively, qualitatively, emotionally, opportunistically, ethically, and/or in some other way?

  o Did your opinion of the risk of Fearlessness Training change due to the success of the first and second pitches?

- What was the effect of cold on the slugger?

- Did new helmets improve the quality or safety of the Fearlessness Training?

- Was the second helmet a redundant safety feature? Did you originally have this opinion or did something change your mind?

- Were traditional procedures for use of the risk matrix properly executed?

  o See NASA's Systems Engineering Guidebook or See the Defense Acquisition University's Risk Management Guideline, etc.

- Assuming that the risk matrix was properly filled out and then updated before each pitch; did the risk matrix adequately convey the risk in order to make a proper decision on whether or not to continue fearlessness training?

- What risk rating, if any, do you suppose would have caused the team manager to stop the Fearlessness Training? Does the manager's use of the risk matrix absolve him of responsibility for the slugger's fate?

Moral

The moral of The Parable of the Fearless Slugger is that there's more to the nature of risk than can be captured in a traditional risk matrix.

The Parable of the Fearless Slugger was inspired by the true stories of the tragic accidents which occurred on NASA's Space Shuttle Challenger, in 1986, and Columbia in 2003, which both involved the acceptance of unacceptable risk. The sentiment of our parable's moral lesson is evident in the conclusions of the final report of the Columbia Accident Investigation Board.

"It is our view that complex systems almost always fail in complex ways, and we believe it would be wrong to reduce the complexities and weaknesses associated with these systems to some simple explanation. Too often, accident investigations blame a failure only on the last step in a complex process, when a more comprehensive understanding of that process could reveal that earlier steps might be equally or even more culpable."[138]

Report of the Columbia Accident Investigation Board

2003

When the causes of a risk are poorly understood we can mistakenly believe that we're facing a relatively inconsequential speculative risk, when we are actually confronting a sophisticated conditional risk. Under these circumstances, instead of an inconsequential loss, we may be exposed to conditions of danger which may be unexpected, uncontrolled, and which may present unlimited potential for loss or harm. For this reason, our first responsibility in managing risk is to understand the nature of the risks we confront with the guidance of the wisdom of a tree.

THE WISDOM OF A TREE

The wisdom of a tree tells us that management of a risk begins with an understanding of the nature of the risk, which must enlighten us and, thereby, influence the implementation of a living strategy that adapts itself to circumventing, counteracting, and triumphing over the risks' adverse effects for the duration of the risks' life cycle.

The wisdom of a tree introduces three of the Process Enrichment philosophy's unique principles regarding risk management.

1.   To properly manage a risk, a risk manager must first recognize the nature of the risk.

---

[138] Report of Columbia Accident Investigation Board (Hereafter referred to as the CAIB), Volume I, Aug. 2003, p. 6.

In the nature of every risk, which is shaped by the risks' strategic causes and by the evolving circumstances of the risks' governing strategy, we find a full lifecycle of phases during which risk transforms for the better or for the worse as the ability to successfully enact an intended strategy changes. Second only to a risks' influence, a risks transformation over time is its most predominant characteristic. The way a risk behaves, as a consequence of trends in a strategy or in response to external conditions and events, and how a risk culminates, either in adverse consequences or in dissolution, are transformations that reveal a risks nature. By understanding the nature of a risk we gain the ability to use our expectations for the risks' transformative behavior over its life cycle, and our awareness of the conditions that emerge, as sources of predictive signals which enable us to craft proper defensive and corrective actions that minimize the detrimental effect of risk and improve the prospects for successful performance of our intended strategy.

This lifecycle conception of risk contrasts sharply with traditional risk management methodologies. If, as in traditional methodologies, we begin risk assessment by over-simplifying risks and consider risks to only be a factor of a probability and a consequence, and we then over-quantify risks, reducing them to the value of a risk exposure, (the product of the probability and the consequence of the risk) we succeed in reducing risks to an easily managed number, but we strip each risk of most of its managerially useful information. In consequence, using traditional risk management techniques, the analysis and response to the risk may not account for and may, therefore, encourage risk managers to disregard unfavorable risk trends or the full impact of the risk in each of quality's six Themes of Performance, and, therefore, on the ability to successfully perform the intended strategy. Under these circumstances, risk management can easily devolve into a process in which, no matter how much manpower is invested in risk management, the proper execution of traditional risk management methods is likely to result in managers unknowingly accepting complex risks and waiting for undesirable consequences to come about. Even worse, since traditional risk management techniques consider only one risk at a time, and do not explicitly consider a creative efforts' system of multiple risks simultaneously, traditional risk management methods have a significantly diminished capability to serve in improving the strategy for a creative effort. These weaknesses in the philosophy and methodology of traditional risk management are, themselves, a significant risk since they may actually increase, rather than decrease, the prospects for the worst manifestations of risks' adversity to occur. In contrast, the nature of a risk holds the enlightening information from which managerially useful information can be gleaned in order to properly respond to risk, secure successful strategic performance, and prevent real-world dangers.

2.   The enlightening experience of risk assessment must serve to produce the wisdom and strategic communication necessary to influence strategy and quality improvement which minimizes risk.

Risk is an influence that, when adequately understood and effectively communicated to decision makers, can motivate transformational change towards quality excellence by bringing about a realization of the necessity to improve performance in order to meet strategic goals. In the wisdom that risk conveys we gain influential evidence that a significant part of the targeted customer groups' strategy cannot be performed successfully, or as successfully as intended, in order to satisfy customers' desires. For this reason, the ability to understand the wisdom risk conveys has a direct relationship to an organizations' ability to compete and survive in a world where intelligent customers can and will continually pursue better alternatives.

3.   Every significant risk requires a managed, living, strategy that adapts itself to the dynamics of circumventing, counteracting and triumphing over the risks' adverse effects throughout the risks' lifecycle.

The best way to manage risk in a creative effort is to consider how risk, in each of quality's six Themes of Performance, can be minimized, from the very beginning, in the design of a product and service and, also, in the conditions and design of the system of product and service development and delivery. This is true because if the inherent risks of the initial strategy are found to be unacceptable, they may cause immediate negative effects and may also be difficult and expensive to correct later on. For risks which cannot or will not be effectively removed through achievement of performance quality excellence, in the early stages of strategic design, a goal of risk management must be to develop an intelligently guided or "living" risk management strategy to circumvent or counteract the adverse effects of these risks as they transform over the course of their lifecycles. Implementing a living strategy is necessary because strategies containing sophisticated risks tend to change dramatically and in ways such that their risks can acquire an even more undesirable potential over time.

To respond to sophisticated risks, especially those found in competitive or adversarial warfare, where adverse intellects are actively shaping and reshaping their strategies to counteract each other, our best approach is to capture and convey the true nature of the risk, how it transforms over time, and how consequences in our ability to attain the desired incentives of our strategies change over time. As we will soon see, these considerations are in the province of risk modeling and management using Saile charts. Even for placid risks, which do not have the dynamics of competitive warfare, Saile charts provide the managerially useful ability to assess and communicate which Themes of Performance are being affected by risk. This provides a realistic context to view the strategic effect of each risk and serves the purpose of managing risk – to expose risk's influence and energize decision makers to enact beneficial strategic improvement.

In commerce, when the nature of risk is understood and minimization of undesirable risk is a prominent consideration, for both producers and acquirers of creative effort, society benefits along with the success of these business enterprises. As any wise tree would tell you, if it could, when organizations triumph over risk, in a very real sense, we all do.

The Nature of Risk

Our rightful intention, in confronting the undesirable risks which affect our strategies, is to develop an appropriate strategy to circumvent or counteract the risk, for the duration of the risks' life cycle, in order to improve the performance of our strategies and triumph over the risk.

A Risks' Lifecycle

Since the Process Enrichment philosophy is almost certainly the first philosophy to attribute a life cycle to risk, it will, perforce, be necessary to introduce a new naming convention for some elements of a risks lifecycle. Hopefully, these new names for risk elements will be found appropriate. An overview of a risks' life cycle, is shown in table 4-1, along with examples of effects of risk that are to be expected and patterns of risk behavior which, in some cases, may predict risks' future behavior and effects.

Phases of a Risks' Life-cycle					
Strategic Conception			Metamorphosis		Reincarnation
Causes of Risk	Condition at Initiation	Routine Active Behavior	Resolution Syndrome	Resolving Condition	Customer Borne Residual Risk

Speculative Causes
▲ Gamble or investment
▲ Innovation
▲ Planned obsolescence
Conditional Causes
▲ Deviation from quality excellence
▲ environment
▲ Supplier dependency
▲ Capability impairment
▲ Complexity
▲ Imprecise estimation
▲ Deficient resources
	▲ Initial risk is understood to some extent and accepted
	▲ Initial strategic intentions (goals) are accepted
	▲ Expectations are set
		▲ Consistent
		▲ Trending
		  o Linear or non-linear (i.e. exponential) increase or decrease
		  o Seasonal variation or oscillation in regular or irregular phases
		▲ Indeterminate
		▲ Impacted by an adverse intellect
		▲ Combination
		▲ None – spontaneous resolution
			▲ Culmination
			▲ Trend / Accelerated Trend
			▲ A critical decision point arises
			▲ An event in a related strategy occurs
			▲ Stepwise increase or decline in the potential for a desirable or undesirable resolution
			▲ No advanced notice or signal
				▲ Profit
				▲ Breakeven
				▲ Loss
				▲ Improvement
				▲ Damage
				▲ Injury or death
				▲ Risk is transferred
				▲ Risk is transformed
				▲ Disposal / Risk is concluded
					▲ Full Value realized
					▲ Un-sustainability
					▲ Unreliability
					▲ Compromised utility
					▲ Latent defects
					▲ Accidental property damage or destruction
					▲ Injury or death
					▲ Disposal / Risk is concluded

Risk Initiation The value of pursuit of a strategy is qualitatively and quantitatively assessed yielding a decision to pursue or not to pursue the strategy given the incentives, conditions, creative effort, resources and cost, schedule and necessary level of quality involved – all of which may vary or deviate from expectations over time. Pursuit of the strategy initiates risk.		**Risk Effects During Routine Active Behavior** ▲ Profit improves or degrades ▲ Value improves or degrades ▲ Performance is acceptable or is unacceptable ▲ There is potential for variation or deviation from quality excellence	**Indicators Predicting Entry into a Risks' Resolution Syndrome** ▲ Events coincide to bring about resolution ▲ A performance trend continues to resolution ▲ A pass or fail event occurs ▲ Alternatives arise requiring a decision ▲ External events or intellect alter risk conditions or incentivize early resolution	**Organizational Impact** ▲ Mission success or failure ▲ Market improvement or decline ▲ Competitor's risk is relatively increased ▲ Reduced Competition	**Market Impact** ▲ Society benefits ▲ Competing products and services are displaced ▲ Competition retreats or heats up ▲ Laws are passed to help or curtail the strategy

Table 4.1: The phases of a risks' life-cycle, pattern indicators, and examples of the effects of risk over a risks' life-cycle.

Any risks' lifecycle can be specified by 3 major phases. In the first phase, "Strategic Conception," risk is created by a strategy whose elements contain the risk. The second phase, during which a risk can transform over time, for better or for worse, is a risks' "Metamorphosis" phase. The third and last phase is one in which a risk has been transferred to another party, as it is, for example, when we buy car insurance or when we sell a product to a customer. Under these circumstances, the risk can, once again, have an enduring life cycle, under the strategy of the risks' new owners. This is a phase called "Reincarnation."

Risks' Strategic Conception Phase

> ### THE SAILE PRINCIPLE
>
> *The root cause of any risk is that the strategy, governing the performance affected by the risk, deviates from a strategy of quality excellence.*

The Saile Principle is a formal expression of the principle underlying the methodology of Saile charting which was introduced in chapter 3. As you may recall, our discussion about necessary improvement and risk was pictured in chapter 3's Saile Charts in Figures 3-1, and 3-2. As those Saile charts indicated, we can visually describe a risk or a system of risks on a Saile Chart as the deviation or deviations of actual performance, from the Decisive Points, strategically defined to indicate quality excellence, in each Theme of Performance. Under competitive circumstances, we can also look at risk as a relative deficiency in quality in one or more of quality's six Themes of Performance. The Saile principle conveys the idea that risks do not stand independent of the strategy in which they are rooted and, also, that risks may broadly impact an entire strategy rather than only an isolated capability. This implies that poor performance caused by risk can have an effect like a series of dominoes – if a risk leads to poor performance in one Theme of Performance, another Theme may also perform poorly or an entire strategy may perform poorly or may collapse in failure if the risk of poor performance of the broader, interconnected, strategy is not adequately considered and addressed.

Causes of Risk

Our definition of risk reveals that all risks are caused by a strategy which deviates from quality excellence, and, further, is inhibited from transforming to quality excellence by either an incentive or a condition which is inherent in the nature of the strategy. If a strategy's incentive is the dominant driver of the deviation from quality excellence, as it is in a gamble or an investment, then the strategy contains predominantly speculative risk. If the conditions under which the strategy is to be executed are the dominant driver of the deviation from quality excellence, as they are when a strategy uses poor quality resources, poor capabilities, or even a lack of knowledge of the customers' strategic needs, then the risk can be recognized as a predominantly conditional risk. In one or more of the elements of strategy, from its incentives, motives, intentions and behaviors using resources, we may have deviation from quality excellence in one or more of the Themes of Performance which cause risk. The significance of understanding the causative factors of risk is that this knowledge helps us to unravel the mystery of a risk, identify and predict where and in what way performance is or will be affected, so that this knowledge can be enlightening and useful in pursuit of ascendant quality.

A Risk's Condition at Initiation

Acquisition and, also, creative effort occur under favorable or unfavorable circumstances as the conditions of the market of the day dictate. Customers are affected by imperfect information and, also, imperfect capabilities to evaluate the creative effort that is available to them in the marketplace. Suppliers are similarly affected by an imperfect capability to understand the strategies that drive customer desires. Sometimes it's best to either wait or, alternatively, to promptly initiate a strategy to take advantage of favorable circumstances. Conditions of high or low resource prices, excessive or slow customer demand, and other factors can affect a strategy's risk from the very beginning. Since these initial conditions of risk may not be obvious, yet they may significantly affect the level of success, it can be worthwhile to consider them. Risk is always relevant to the present, although its causes may evolve and become more or less evident and more or less problematic in the future.

Risks' Metamorphosis Phase

A risks' Metamorphosis phase consists of three distinct episodes. The first episode is the risks' "Routine Active Behavior." This episode varies in duration but is the calm before the storm of many risks. The second episode is the risks' "Resolution Syndrome." The risk, having persisted through its Routine Active Behavior episode, without being remediated, is now in an episode where it changes its character by culminating in a trend, an event, a decision or the like, as it progresses towards the final episode, which is the risks' "Resolving Condition." In its Resolving Condition, a risk yields an outcome such as a profit or loss, successful or unsuccessful performance, physical damage or an injury.

The influence and effects of a risk are usually not constant. Risks can become more or less able to produce adverse effects over time as mitigation steps, conditional changes, or incentive changes occur. As the actual or potential adverse effects of a risk increase or decrease, the influence to respond and mitigate the risk changes. For this reason, large organizations are inclined to carefully track the progression of risk to make sure that a strategic correction or mitigation occurs before risks that are acceptable morph into risks that are unacceptable. Saile Charts offer an excellent visual system for communicating the progression of one or more risks over time to show the changing effect of a system of risk on the ability to perform the intended strategy. Examples of the use of Saile Charts to convey the anticipated transformation of risk over time were presented in chapter 3's Figures 3-5 and 3-7.

Routine Active Behavior

Although it is challenging enough to consider the risk in a single strategy, it is frequently the case that several strategies may serve a significant organizing strategy at the center of gravity. In these cases, where, for example, an organization has multiple product lines or significantly different types of customers to serve, using some of the same resources, these strategies may interfere with each other at seemingly random times. This can cause new risks and can exacerbate existing risks as these strategies compete for resources and as some strategies dominate or are dominated by others. Strategies that are sometimes at cross purposes can act together to create a complex decision environment permeated by imperfect knowledge of current or evolving conditions. Although we should remain cognizant of the interacting strategies that may at any time act in cross purposes, events in simultaneous strategies may vary in timing and degree and in unforeseen ways. For these reasons, we shouldn't expect risk to be a consistent, always acceptable and predictable, influence – but we should predict and manage the conditions that we can.

Resolution Syndrome

In a risks' resolution syndrome, a risk acquires the characteristics that will culminate in the risks' resolving condition. This acquisition of one or more characteristics is composed of signs that risk managers should be looking for to predict the near future outcome of a risk. Discovery of the initiation of the risks' Resolution Syndrome also indicates that efforts to prevent the Resolution Syndrome from occurring have failed. The preventive efforts, if any, may also have lost their relevancy and may now be counterproductive. The risk management strategy must be reconsidered and must adapt as necessary to the exigencies of this new phase of the risk's lifecycle in order to bring about the most favorable outcome.

Indicators Predicting Entry Into a Risks' Resolution Syndrome

There are risks, which, like bad brakes on a car, can give out a screech to tell the driver that what has previously been accepted as Routine Active Behavior has entered a stage where it has now taken a turn for the worse. The sound of screechy brakes signals that, having ignored the mechanics warnings during routine service, the brake pads have completely worn through and the mechanic must now do the more expensive job of "cutting the rotors" to give the new brake pads a good surface to grab on to. When, as in most creative processes, there isn't a natural screech to signify entry into a risks resolution syndrome, we may contrive signals based on physical changes, process performance changes, or other measurements that can do the screeching for our creative effort. For example, poor sales, loss of key customers, late deliveries, cost overruns, new competitors entering the market, changes in financial ratios, a projects' lengthening critical path, customer complaints, minimum inventory being reached, etc.

Some risks that are an integral part of a product; as a consequence of its design like, for example, the plastic that covers a car's bumpers, will not have a Resolution Syndrome. The plastic will simply crack on impact.

Resolving Condition

A risks' resolving condition reveals the extent to which strategic incentives for the creative effort have been attained (i.e. profit, loss, improvement, etc.) so far.

Risks' Reincarnation Phase

In creative effort, when risks are not remediated, the risks are likely to travel along with the creative effort to customers and then to the customers of customers. The ability of risk to be transferred to a new owner, while still retaining its risky nature, means that risk can, in a sense, be reincarnated to go through its lifecycle again. Like a vampire, risk can take on a new appearance, lay dormant, or turn violent with each new life it takes and, like stopping a vampire, the only way to stop a risk is in the proper dissolution of the strategy and the products and services possessing the risk. This is why society, from here to Transylvania, is better off when creative effort bears a minimum of risk.

Customer Borne Residual Risk

When a poorly designed product is sold, customers receive a product that they believe may serve their purposes, but they also receive something they almost never want – risk. Risk often reduces the value of a product and service to customers because it is symptomatic of a reduced ability of the product or service to perform their intended strategies or to sustain performance of their intended strategies reliably. When customers are aware of the risks of alternative products and services, the alternative of lowest risk will be found most attractive and valuable – although customers may choose to take their chances on riskier products and services if there is a discount sufficient to compensate for the risk.

Organizational Impact

There is a wide range of very positive to very negative outcomes that can be expected from the resolution of a risk. For example, an organization performing a software development project that is high risk, due to the complexity or creativity involved, may charge a premium for the effort and, consequently, may receive better profit to the extent that the project successfully manages these risks.

When a product and service are made available to a market they become an alternative among products and services offered by competing firms. This may cause a reaction among competitors ranging from mild price adjustments to competitive warfare. If there are no alternatives, and the offering is needed, the creative effort will be of benefit to society.

To recap, we can readily understand that the root cause of a gambling loss may be inherent in the strategy of gambling because we are aware that casinos and other gambling venues typically stack the odds in their own favor. As a discussion earlier in this chapter illustrated, once a gambler drops a quarter into a slot machine, the speculative gambling strategy has initiated as has the uneventful "Routine Active Behavior" of the risk. It's a matter of pulling the handle on the one armed bandit to enter the risks Resolution Syndrome. The risks' Resolution Syndrome then concludes in one of several ways. The quarter could be lost or the gambler may have won $50,000. If the quarter is lost, this would resolve a rather harmless risk, however, if the gambler won $50,000, the risk reincarnates in a new form. Hopefully, in its new form, the risk yields a benefit to the gambler, but, as some lottery winners can attest, sometimes winning is when lifestyle changes that the winner cannot adequately cope with bring adversity. By identifying the root cause of a risk as a strategic element we now have a system of elements we can look at as elements that can potentially be improved assuming we wish to mitigate or remediate the risk to improve our chances of beating the odds.

Our effort to understand risks' life cycle provides the preparation necessary to develop a strategy to circumvent, counteract and triumph over risk.

To circumvent, counteract, and triumph over any risk, to the extent that it's possible to do so, requires determination:

1) Determine which strategy is the governing strategy containing elements which create the risk.

2) Determine the incentives and expectations regarding them, the motives, intentions and the abilities to perform, in each Theme of Performance, which create the standard of quality excellence for the strategy.

3) Determine the actual or expected performance in each Theme of Performance.

4) Determine the risks' lifecycle based on the strategy and how it is expected to change over time given the potential for changes in incentives or conditions.

5) Determine which measurements can be implemented to signal the risks' phase or episode transitions

6) If the strategy creating risk is within your organization, or is within your control, determine how best to:

    a. introduce ascendant quality that raises the ability to perform to the level of quality excellence or beyond or to the intended level of performance in each Theme of Performance;

    b. transfer the risk to a person or organization that improves your or your organizations' ability to attain strategic incentives; or

    c. stop performing the strategy if it can't be made acceptable.

If the strategy creating risk is an adversary's strategy, determine how best to:

    d. increase the adversary's strategy's risk of poor performance in each Theme of Performance to the extent that the adversary's strategy is nearly impossible to perform with any success; and / or

    e. make success of the strategy serve your own strategy.

7) Determine the quality of the risk management strategy in terms of the improvement of the ability of customers to perform their desired strategy and attain its incentives.

Exercise:

Composing a Risk Management Strategy to Circumvent and Counteract the Risk of Improvised Explosive Devices Emplaced in Roadways

Among the deadliest and most injurious strategies of warfare facing U.S. and other coalition forces, in the recent wars in Iraq and Afghanistan, is the enemy's strategy of attack using Improvised Explosive Devices (IEDs). In Iraq and Afghanistan, building IEDs has been made possible due to the abundant flow of military explosives as well as other materials which can be made to explode. IEDs have proven to be effective for the enemy when they are clandestinely emplaced where coalition forces travel either on foot or in military vehicles. Once an IED is emplaced it can be triggered by movement or by weight, in the same ways as landmines are triggered. Some types of IED can be triggered by wires connected to them, or wirelessly by cell phones or radio signals controlled by the enemy. IEDs are used in and around highly trafficked areas, such as roadways, in order to maximize the potential for destructiveness and to improve the enemy's ability to select from valuable targets. The impact of IEDs, on coalition strategy, is to increase the risk of poor performance to coalition forces and civilians as they try to exercise the freedom of movement necessary to support a viable and progressive contemporary civilization.

In competitive warfare, as well as in military warfare, as the warfare degenerates from cold to hot, each adversary is effective to the extent that they are able to increase the risk of poor performance for their adversary, particularly in the central strategies possessing the gravitational energy which organizes their behavior. Unless a settlement or a "modus vivendi" is reached among adversaries, the increase of risk directed by each adversary against each other must continue until one party or the other no longer has the will or ability to continue the struggle. The Process Enrichment Philosophy expresses the relationship of relative risk, among adversaries, to winning in competitive warfare, in The Fundamental Law of Competitive Warfare.

> *The basis for winning competitive warfare consists, essentially, in the achievement of one condition – competitors with a relatively high risk of poor performance.*

A condition of risk, and, especially, a condition of excessive risk, which is caused when a vital strategy is compelled to deviate from quality excellence, through the ingenuity and manipulation of an adversary, is the only condition, other than a condition of self-inflicted risk, that is, ultimately, decisive in any form of competitive warfare. The greatest generals in history are often noted for their victories in using a small force to conquer a larger or better equipped force. A general's greatness is typically recognized when they take on a great military risk and they succeed – or, more correctly, they increase the risk of poor performance of the larger, more powerful, opposing force to the point where the larger force becomes unable or unwilling to continue resistance. An important part of achieving this reversal of risk, and breaking the enemy's resistance, is the enemy's enlightened knowledge that, even if they could continue to resist, their strategies of resistance bear so much risk that they offer little ability to achieve success. In competitive warfare it is the audacious and clever ability to take the energy out of the enemy's strategy, by increasing their risk, which is the essence of great generalship.

In the enemy's use of IEDs, in Iraq and Afghanistan, it is evident that they would like to increase the risk to freedom of movement because it is this freedom which enables the coalition troops and the civilian population to be supplied with resources and to conduct commerce and the other essentials of a progressive and properly functioning society. These are activities which are contrary to the enemy's desires and to their hopes of seizing political leadership in Iraq and Afghanistan when coalition forces leave. In response, coalition forces have actively sought innovative and effective ways to circumvent, counteract and triumph over the enemy's IED attack strategy by increasing its risk of poor performance. Since IEDs have been used in far too many ways to discuss all of them here, for the purposes of our exercise, we focus only on IEDs embedded in roadways.

Assessment of Current Coalition Strategy

A strategy of attack using IEDs embedded in roadways requires a lifecycle with a sequence of three key initiatives – **building**, **emplacing**, and **detonating** the IEDs. For each of these three initiatives, we can identify coalition effort to increase the enemy's risk.

▶ Increasing the risk of **building** IEDs has been pursued through effort to make the bomb makers, materials, and machines necessary to build IEDs inaccessible to the enemy to the extent possible. Part of this effort involves searching houses of suspected enemies and, as necessary, making arrests and confiscating bomb making materials. It has also proven necessary to monitor border crossings in order to intercept arms and explosives that are being smuggled in. To prevent IEDs from being brought through security checkpoints, equipment that can detect chemicals used in explosives is used as cars and people are searched. Communication among suspected enemies is being monitored in an attempt to discover planned enemy activity. Drone attacks and other military attacks serve to reduce the enemy's quality of life to the extent that recruiting new enemies can be more difficult. To some extent, strategic communication is being used to ostracize the enemy due to the deaths and injuries of civilians. Improvement in personal freedoms, civil rights, and effective utilities producing electricity and clean water, schools for all, and democracy have helped to win over some, but not all, of the enemy. Since the enemy can hide among the civilian population, and can build IEDs in any home, it has proven difficult to stop the building of IEDs.

▶ Increasing the risk of **emplacing** IEDs has been achieved by monitoring roadways and recruiting security forces to assist in the effort. Unmanned drones, manned aircraft, and spy satellites also enable coalition forces to observe enemy activity on roadways. Once IEDs have been emplaced, to detect them, soldiers regularly sweep roadways with metal detectors similar to those used on the beach to find lost jewelry. To destroy IEDs which have been emplaced but haven't been detected, heavy trucks are used in front of convoys to trigger pressure sensitive IEDs as safely as possible. These techniques are effective; however, the time and resources necessary for them limits their ability to be used on the many roadways where restoring safety is necessary. Since the process of emplacing IEDs takes a considerable amount of time, many of the enemy have been captured or killed in the process – but not enough to stop the process.

▶ Increasing the risk of **detonating** IEDs is difficult because, once IEDs have been emplaced, they are ready for detonation at a moment's notice. At the detonation stage, the enemy can operate relatively safely and only obvious suspicious behavior can alert coalition forces to their danger. After the IED explodes it may be possible for forensic scientists to determine the source of the IEDs components or the bomb maker's identity and to try to eliminate these resources.

The enemy's governing strategy of attack using IED's embedded in roadways has, already, been circumvented and counteracted, with good effect, using many innovative tools and techniques which have, undoubtedly, served to increase the enemy's risk of poor performance. Although an intense focus on innovative creative effort, from across the world, has been brought to bear in the effort to respond to IEDs, the civilized world is not content with current results because the menace of IEDs persists with terrible and unacceptable effect.

The success of our effort to improve the design of the coalition's risk management strategy, in accordance with the Fundamental Law of Competitive Warfare, will be measured by the reduction of risk to the coalition's strategy of exercising freedom of movement in Iraq and Afghanistan.

Assessment of Current Enemy IED Attack Strategy

Whether we seek to improve our own strategy, or to counteract or circumvent an adversary's strategy, strategic analysis is crucial because there is the potential within every element of strategy to stimulate creation of the valuable insight that can make improvement possible. Assessing performance within the Themes of Performance, for the enemy's IED attack strategy's three key initiatives – **build**, **emplace**, and **detonate** – and an assessment of performance of subordinate strategies supporting these initiatives are described below and presented on the left side of Figure 4-3.

• In order to **build** an IED, considering the Acceptance Theme of Performance, the enemy has to have a strategy to obtain the necessary components and other resources that will comprise the IED. Considering the Efficacy Theme of Performance, the enemy needs a strategy to give potency to the IEDs explosion.

- In order to **emplace** the IED, considering the Acceptance Theme of Performance, the emplaced IED should be properly armed and ready for detonation. Considering the Efficacy Theme of Performance, the IED should be as undetectable as possible. This is brought about by reusing the asphalt, cement or dirt of the original roadway. Restoring the roadway to the original condition helps prevent coalition forces from discovering and neutralizing the IED before it is used.

- Lastly, in order to **detonate** the IED, considering the Acceptance Theme of Performance, the enemy would like the IED to be capable of being detonated at an appropriate opportunity. Considering the Efficacy Theme of Performance, the enemy would like to detonate the IED when desirable targets can be struck.

It wasn't necessary to carry the Thematic analysis of enemy strategy higher than the Efficacy Theme of Performance for a very good reason; the analysis has already uncovered a critical vulnerability in the IED Attack strategy.

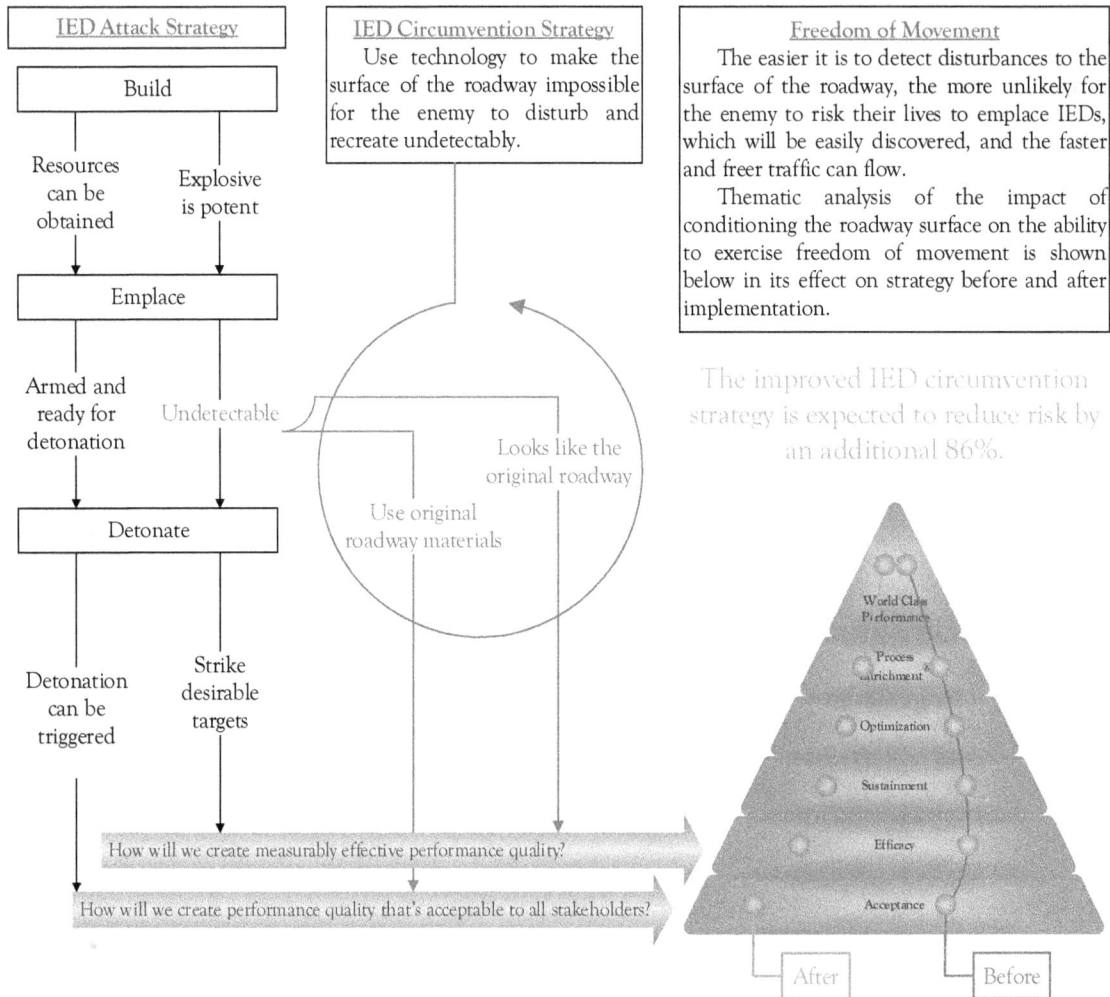

Figure 4-3: Strategic analysis of the enemy's IED attack strategy reveals a critical vulnerability in the subordinate strategy of making the IED undetectable, which is the strategy that gives emplacement of the IED its efficacy. An IED circumvention strategy is shown to reduce Thematic Risk Exposure to freedom of movement on roadways by 86% over current practices.

An overview of the elements of coalition, as well as enemy, strategies is shown in Figure 4-3. A critical vulnerability in the enemy's IED attack strategy arises from the fact that in order to perform the "Emplace" initiative, with efficacy, an IED needs to be embedded in the roadway in a manner that makes the emplaced IED "Undetectable." A Thematic analysis of the strategy to make emplacement of an IED undetectable is shown in the center of Figure 4-4. We can see that the "undetectable" strategy for the IED will meet expectations in the Acceptance Theme of Performance if the "original roadway materials" are used and it will meet expectations for the Efficacy Theme of Performance if the roadway over the IED "looks like the original roadway." The enemy has had great success emplacing IEDs because concrete or asphalt for the roadway surface is generic and easy to obtain. The enemy can either bring along standard grey cement or black asphalt or simply reuse the cement or asphalt dug up in order to emplace the IED. Using a trowel, or some flat tool, bombers have been able to camouflage their IEDs with the same ease that a pothole in the roadway is filled in. This is far too easy for the enemy to do – but coalition forces could easily make this strategy far more difficult and perhaps impossible to perform.

IED Circumvention Strategy

There are probably dozens of methods which could be used to make the surface of any roadway impossible for the enemy to recreate. One can easily imagine painting the roadway a unique color, engraving or chemically etching a complex design or, perhaps, a computer generated fractal or kaleidoscopic design on the roadway surface. Unlike concrete and asphalt surfaces which can be recreated easily, if a hole is dug through a complex pattern on the roadway surface it can be impossible to recreate the design and, also, impossible to hide the edges of the hole, without the same technology as the road builder. If the enemy cannot recreate the surface of the roadway, after implanting an IED, then they cannot make their IEDs undetectable. If the strategy of making the IEDs undetectable fails, the entire IED attack strategy, essentially, fails because coalition forces will readily dispose of IEDs they can detect. The ability to see where IEDs are emplaced removes most of their efficacy – there is still significant danger in the process of destroying them. Another advantage of incorporating technology in the roadway surfaces is that there will be more and more high quality, improved, roadways; also, as the technology for surfacing the roads improves, traffic can move faster and, perhaps, the roadways can be scanned quickly using drones flying over them.

Under conditions where technology in the roadway makes it impossible to emplace IEDs undetectably, the enemy may become unwilling to risk their lives or risk imprisonment to carry out an IED attack strategy. In the counteractive strategy we developed, using strategic analysis, we have met the condition that is the basis for success, as expressed in the Fundamental Law of Competitive Warfare, by increasing the enemy's relative risk and improving freedom of movement. Coalition forces should seek to continuously increase the risk of poor performance for any strategy that the enemy seeks to exercise – except for their strategies leading to surrender or cessation of hostilities – which should be helped along.

Thematic Risk Exposure Evaluation (TREE)

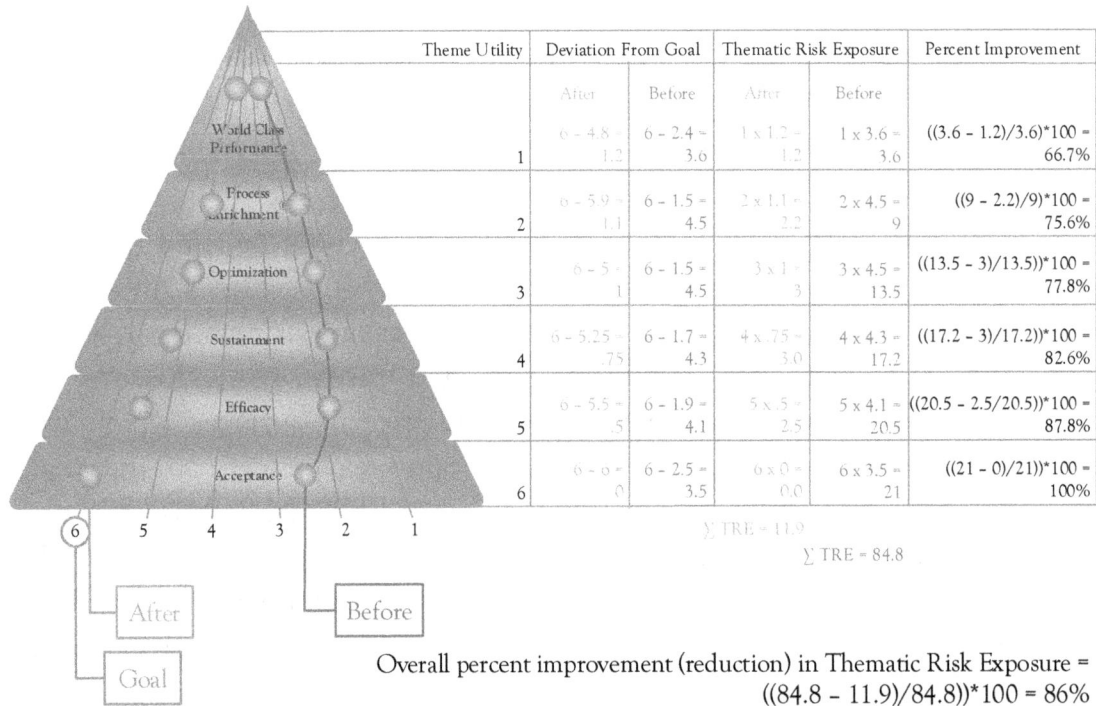

Theme (Utility)	Deviation From Goal		Thematic Risk Exposure		Percent Improvement
	After	Before	After	Before	
World Class Performance (1)	6 - 4.8 = 1.2	6 - 2.4 = 3.6	1 x 1.2 = 1.2	1 x 3.6 = 3.6	((3.6 - 1.2)/3.6)*100 = 66.7%
Process Enrichment (2)	6 - 5.9 = 1.1	6 - 1.5 = 4.5	2 x 1.1 = 2.2	2 x 4.5 = 9	((9 - 2.2)/9)*100 = 75.6%
Optimization (3)	6 - 5 = 1	6 - 1.5 = 4.5	3 x 1 = 3	3 x 4.5 = 13.5	((13.5 - 3)/13.5)*100 = 77.8%
Sustainment (4)	6 - 5.25 = .75	6 - 1.7 = 4.3	4 x .75 = 3.0	4 x 4.3 = 17.2	((17.2 - 3)/17.2)*100 = 82.6%
Efficacy (5)	6 - 5.5 = .5	6 - 1.9 = 4.1	5 x .5 = 2.5	5 x 4.1 = 20.5	((20.5 - 2.5/20.5))*100 = 87.8%
Acceptance (6)	6 - 6 = 0	6 - 2.5 = 3.5	6 x 0 = 0.0	6 x 3.5 = 21	((21 - 0)/21)*100 = 100%

$\sum$ TRE = 11.9

$\sum$ TRE = 84.8

Overall percent improvement (reduction) in Thematic Risk Exposure = ((84.8 - 11.9)/84.8))*100 = 86%

Figure 4-4: Calculation of the expected Thematic Risk Exposure (TRE) improvement after implementation of the IED circumvention strategy.

The strategic analysis, as shown in Figure 4-3, is the basis for the Thematic Risk Exposure Evaluation (TREE) shown in Figure 4-4. The acronym TREE will, hopefully, remind strategists of the wisdom of a tree. It was just such wisdom which has shown us that to cope with the risk presented by the enemy's IED attack strategy we needed to be enlightened by our understanding of the nature of the risk through analysis of the elements of strategy. In this way, we were able to identify a critical vulnerability, which we could exploit, leading to an 86% reduction in Thematic Risk Exposure. The value of this improvement, in reduced property damage and casualties and in the gain of freedom of movement, appears to be compelling.

Admittedly, if someone brainstorms intently, and long enough, they may arrive at the same innovative ideas that we may discover, relatively quickly, through application of the Process Enrichment philosophy; however, we may find that a philosophical approach may be a better way to ensure that the brain, while brainstorming, has a roadmap to follow where the well trodden path is highly likely to expose a feasible and prefferred solution if there is one. For example, the IED circumvention strategy we just developed presents a solution to the enemy's IED attack strategy which has eluded the best brainstorming efforts of tens of thousands of soldiers, and dozens of the worlds best think tanks, who have spent many years grappling with this deadly and intractible problem.

We are now ready to enjoy exploration of a new case study which, like those before, is steeped in the flow of human trajedy as well as in the ascendancy of the human spirit and the human capacity to rise above and beyond all that challenges us.

Case Study:

NASA's Space Shuttle Program and the Fearlessness Effect

The challenge of creating an organizational culture that reacts to risk

- How will we avoid the hazards of risk when creating safety-critical products and services – where failure is not an option?

Introduction

In safety-critical applications, where products and services bear risk with the potential for injury or for deadly consequences, organizational leaders must be aware that the risks are not found only in the technology of the system produced, but are also found in the technology of the organizational processes and approaches to achieving quality excellence as well, because it is with these processes and approaches that safety-critical products and services are declared safe and "good to go." Complex organizations creating complex, state of the art, safety-critical products and services are especially vulnerable to habits and conventional wisdom that is obsolete with respect to the challenges of their new creative effort. Where management fails to adapt the technology of their organizational methods for creating quality and safety, as necessary to meet the new challenges brought out by the technology and services their organization is producing, risk can masquerade in the façade of quality, defeating an organization's protective measures and removing people's protective fear, to create the Fearlessness Effect, which is to only be unmasked in the culmination of a tragedy.

Prologue

As organizations coordinate their effort around a strategy serving as a powerful Center of Gravity, the talents and efforts of craftsmen, scientists, and various professionals can, in Coascendance, build great things that elevate the pride of mankind everywhere. Creative effort to attain great rewards often presents great risks as well. A sophisticated organization must be capable of responding to the incentives and conditions that cause risk and must also be aware of the incentives and conditions that lead risks to be suppressed and hidden in major programs. Technical fallibility, human error, and fate are often more easily blamed when failures occur then are lapses in the capability to adequately understand and manage quality and risk. Often, in consequence of the many strengths an organization has, the organization can become overconfident and can take management of quality and risk for granted and an organization can become dangerously Fearless.

NASA's Space Shuttle Program and the Fearlessness Effect

Considering the complexity of the Space Shuttle System (SSS) and the extraordinary number of successful missions the SSS vehicles have completed, it would be reasonable to consider the SSS among mankind's greatest creative efforts – even though two catastrophic accidents occurred over the program's 30 year run. However disquieting it may be to recall these two accidents, which brought down NASA's Space Shuttles Challenger, in 1986, and Columbia, in 2003, it is necessary that we explore these incidents to expose some important, and, as yet, unlearned lessons that these two seemingly disparate accidents can bring light to.

The losses of Challenger and Columbia were events, for those of us who lived through either of them, which weighed heavily on our hearts, as these missions, for the advancement of science, cost the lives of 14 remarkable people.

Challenger – 1986	Columbia – 2003
Greg Jarvis	Michael P. Anderson
Christa McAuliffe	David M. Brown
Ronald McNair	Kalpana Chawla
Ellison Onizuka	Laurel Blair Salton Clark
Judith Resnik	Rick D. Husband
Francis R. Scobee	William C. McCool
Michael J. Smith	Ilan Ramon

We should also take note that, with varying degrees of risk to their careers, in the events leading up to both Space Shuttle accidents, there were people like John Miller, Roger Boisjoly, and Rodney Rocha who made their best attempts, before catastrophe struck, to convince decision makers that, because of risk, the Space Shuttles were not ready to launch or were in danger. We will never know all of the people who tried and didn't succeed in preventing these two Space

Shuttle accidents or those who succeeded in preventing catastrophe on the 133 other successful Space Shuttle missions – but we can take pride in knowing that they were there, doing what was necessary, and doing it as best they could.

External Tank

Space Shuttle Orbiter

Solid Rocket Boosters

Figure 4-5, NASA's Space Shuttle System (SSS) consisting of the Space Shuttle Orbiter, two Solid Rocket Boosters (SRBs), and a liquid fuel filled External Tank (ET).

NASA's Descriptions of the two Space Shuttle accidents

1986, Challenger

"The explosion occurred 73 seconds into the flight as a result of a leak in one of two Solid Rocket Boosters that ignited the main liquid fuel tank."[139]

2003, Columbia

"As the Space Shuttle lifted off from Kennedy Space Center in Florida on January 16, a small portion of foam broke away from

---

[139] The Crew of the Challenger Shuttle Mission in 1986, NASA. Updated 10/22/2004. Downloaded on 6/29/2014 from http://history.nasa.gov/Biographies/challenger.html

the orange external fuel tank and struck the orbiter's left wing. The resulting damage created a hole in the wing's leading edge, which caused the vehicle to break apart during reentry to Earth's atmosphere on February 1."[140]

Following the Columbia disaster, the Columbia Accident Investigation Board (CAIB) investigating the disaster lamented that "the causes of the institutional failure responsible for Challenger have not been fixed"[141] and that "if these persistent, systemic flaws are not resolved, the scene is set for another accident. Therefore, the recommendations for change are not only for fixing the Shuttle's technical system, but also for fixing each part of the organizational system that produced Columbia's failure."[142]

Attempting to resolve past errors using the same body of knowledge and system of assessment that allowed and enabled the errors to happen in the first place may yield incremental improvements of great value, however, based on the new conceptual ideas that we have worked through in the last few chapters, we may be able to shed new light on the limitations of the old methods and thereby open the path for a new way of thinking about risk and quality and new methods to approach these issues that are so vital to success in safety-critical missions.

Although we can appreciate that working towards the success of a valued strategy is a noble endeavor, we must be cautious so that the power of the desire to attain the incentives of our strategy doesn't blind us to the risks that may come to light as we execute our chosen course of action. An accumulation of flaws and fallacies in the last century's beliefs regarding strategy, quality, and risk can culminate in a distorted conceptual system which can produce what I call the Fearlessness Effect.

---

[140] Space Shuttle Overview: Columbia (OV-102), NASA. Jeanne Ryba, Editor. Updated 4/12/2013. Downloaded on 6/29/2014 from http://www.nasa.gov/centers/kennedy/shuttleoperations/orbiters/columbia_info.html
[141] CAIB, Vol. 1, p.195, retrieved from http://www.nasa.gov/columbia/home/CAIB_Vol1.html
[142] CAIB, Chapter 8, Vol. 1, p.195

The capacity of the Fearlessness Effect to attain its most adverse proportions is enhanced by six fallacious beliefs.

1) The fallacious belief that "quality is conformance to requirements" obstructs adequate inquiry into the ability of creative effort to perform successfully, as intended, in each of quality's six Themes of Performance, thus perpetuating design flaws, invalid or insufficient requirements, and other risks of poor performance including hazards.

2) The fallacious belief that "fixing defects improves quality" obstructs correction of process or product design inadequacies that make creative effort defect prone, risky, unreliable, and perhaps dangerous.

3) The fallacious belief that "measurement and control of variation improves quality" influences decision making to the extent that poor or unacceptable performance yielding deviation from an intended strategy is not properly recognized and is, instead, masked as a form of acceptable performance yielding variation and occasional extreme variation or outliers.

4) The fallacious belief that risk should be defined and managed as a circumstance with a probability of occurrence and a consequence, rather than as an influential circumstance with a lifecycle, causes risks to be accepted and ignored, disregarding trends signifying trouble and suppressing effort to correct the systemic impact of risk which diminishes the value of creative effort.

5) The fallacious belief that better processes make better quality products is relied on to the extent that it obstructs adequate inquiry into whether the organization itself, its processes, products, and services, actually have desirable

quality and value with respect to the strategic needs of customers. (Exception: The only process that will improve quality is a process that shows how to improve quality. Any other process, at best, maintains the quality that was already designed in.)

6) The fallacious belief that it is possible to manage by fact is used to obstruct the use of common sense, intuitive, but sound, reasoning, or the use of professional opinion to enact corrections to a risky strategy.

It is my contention that the best explanation, for the tragic loss of life in both Space Shuttle accidents, is that NASA's organizational culture harbored strategies of operation that were managed under the influence of the Fearlessness Effect. In the following case study, the manner in which the fallacious beliefs of the Fearlessness Effect can operate, with devastating results on an organizational culture, is introduced in a phased prototype. With this prototype, it will be shown that the Fearlessness Effect occurred, in nearly the same manner, on two Space Shuttle accidents separated by seventeen years. It is also my contention that the many commissions investigating the Space Shuttle accidents, which included many PhDs, and at least one Nobel Laureate, incorrectly viewed the organizational aspects of the accidents simply as bad management and mistakes in following proper risk management procedures rather than as the culmination of a fundamentally flawed cluster of fallacious beliefs operating under the overwhelming gravitational influence of a powerful strategy – as defined by the Fearlessness Effect.

The overwhelming gravitational influence of the strategies to proceed with the missions of the Space Shuttles Challenger in 1986 and Columbia in 2003, enhanced by fallacious beliefs, and the increasing willingness to disown responsibility which these fallacious beliefs make possible, were the actual cause of the loss of life in both Space Shuttle accidents because, in the absence of the Fearlessness Effect, the design flaws, which actually caused the accidents, could have been counteracted effectively. In the analysis that follows, evidence that some of NASA's senior managers were afflicted by the Fearlessness Effect, for both Space Shuttle accidents, is presented entirely through the use of independent, well informed, and well-funded investigators, including: The Presidential Commission on the Challenger Accident; The Columbia Accident Investigation Board; The House of Representatives Investigation; The Atlantic

Monthly; The Washington Post; and several NASA and NASA contractor personnel.

In many organizations it may be possible to recognize some or all of the phases of the Fearlessness Effect active to varying extents. The Space Shuttle program was chosen as the defining example of the Fearlessness Effect not because NASA is any worse in managing risk than any other organization; on the contrary, the intention of this study is to demonstrate how even the performance of the world's best managers of risk can be improved by eradicating the Fearlessness Effect.

Phases of the Fearlessness Effect

1- The Strategic Conception Phase
The customers' interests are embodied in product and service requirements.

2- The Recognition of Unacceptable Performance Phase
Unacceptable performance occurs and is observed.

3- The Defect Repair Phase
Defects are repaired, when necessary, following use of a product or service. The process of repairing defective products or services improves to become a smooth operation.

4- The Delusionary Phase
The fallacious beliefs that 1) fixing defects improves quality; 2) improving processes improves quality; or that 3) risks do not have a lifecycle, act to enable responsible parties to form a delusion with which they are psychologically and organizationally enabled to evade correction of product and service designs or conditions yielding poor or unacceptable performance.

5- The Straw Man Phase
A history of lucky performance is set up as the basis for "Straw Man" requirements which are substituted for the customers' actual requirements. In this way, the customer's interests are disembodied from working requirements.

6- The Conformance to Straw Man Requirements Phase
Creativity and creative effort to correct poor or unacceptable performance is inhibited by commitment to the intention to conform to disembodied Straw Man requirements.

7- The Acceptable Risk Phase
Poor or unacceptable performance, which has evaded correction, is masked as "acceptable risk" due to a growing history of luck and the growing history of defects proving to be repairable.

8- The Masking of Deviation as Variation Phase

To preserve the energy of a key strategy, significant deviations in performance, indicative of sophisticated risks with adverse trends, are often mischaracterized as extremes of variation or as "outliers" of a system of normal variation. In this way, the true nature of design or execution risks is masked, and causes of the risks remain, preserving the risk of poor performance.

9- The Fearlessness Phase

Unacceptable performance, bearing risk, masquerades into the whirlwind of acceptable performance, gravitating towards a common strategy, and creates the false impression of energizing the strategy, while it is, instead, harboring a potential for unrestrained adversity.

10- The Black Operations Liberating Black Swans Phase

Management reflexively, and irresistibly, acts to suppress risk warnings in order to defend and sustain a key strategy. The compulsion to avoid unmasking deviation and decisions which created straw man requirements becomes more controlling over behavior than the legitimate responsibility to respond to risk and prevent adversity. In this way, risks with the potential for rare and dangerous "Black Swan Events" are liberated to evolve under inadequate control.

11- The Concluding Phase

The project completes successfully or an accident occurs.

Table 4-1: A phased, lifecycle view, of The Fearlessness Effect.

Accident Summaries:

The generally accepted facts of the two Space Shuttle accidents are presented in the summaries below. We will reconsider these summaries, in view of the theory that officials at NASA were operating under the influence of the Fearlessness Effect. The phased sequence in which the Fearlessness Effect is proposed to have operated at NASA is presented in side by side columns in order to demonstrate that the Fearlessness Effect operated similarly during the time of both Space Shuttle accidents. Extensive use of direct quotations, from original sources, is made to independently validate the presence of thinking that reflects the influence of the Fearlessness Effect. Hopefully, this in-depth analysis will provide new insight to the risk an organizational culture can harbor to the extent that the vicious cycle of the Fearlessness Effect will become vulnerable to our improved knowledge of risk.

Challenger – 1986	Columbia – 2003
To understand the technical cause of the Challenger accident, we need to learn a little rocket science. The part of one of Challenger's Solid Rocket Boosters (SRBs) that failed is a rubber O-ring which is used to create a seal between sections of the SRB which are assembled together. Like	Insulating foam, used to keep the fuel in Columbia's External Tank (ET) extremely cold, is glued to the outside of the ET. The foam tends to shred off under the tremendous air pressure and friction created as the Space Shuttle accelerates following launch. NASA believes that "trapped air

217

connecting several garden hoses together to make a really long hose, if there isn't a good fitting washer or O-ring in the connection between sections of hose the water will leak out. The difference between the O-ring in a garden hose and the O-rings on the Challenger's SRBs is that, the O-rings are much bigger and the sections of the SRB connected by the O-rings contain hundreds of thousands of pounds of solid rocket fuel which, when ignited, creates combustion gasses at over a thousand degrees. Any escape of hot combustion gases, through the O-rings, was likely, one way or another, to lead to the Space Shuttle's destruction. Another difference in O-rings is that the SRB's O-rings were not merely under compression as in a garden hose, the SRB's O-rings had to flex rapidly to match the convulsive expansion and contraction and wobbling of the SRB during the furious conditions of the 4.5 million pound Space Shuttle's blast off and propulsion into space towards an escape velocity faster than 17,000 miles per hour.

About two minutes after a successful launch of the Space Shuttle, the SRBs are jettisoned and they parachute down to the sea where they are recovered. The SRBs then receive a post flight inspection and are refurbished and made ready for use on the next Space Shuttle flight. During post flight inspection it was discovered that some of the O-rings, which are used in pairs for each joint on the SRBs, showed evidence of significant wear or erosion damage caused by combustion gasses hitting them. Evidence was also found of gasses leaking through the primary O-ring. The second O-ring was believed to be a back-up for the primary O-ring but it was found that both O-rings can fail from the same event and they were, thus,

pockets which expand due to altitude and aerothermal heating" were the cause of small pieces and also small chunks of insulation foam debris falling off in the seconds following launch. Requirements for the ET specifically prohibited debris from the ET from hitting the Orbiter, but this happened on every flight and damaged many of the tiles in the Orbiter's Thermal Protection System. These tiles had to be replaced prior to turnaround for the next Shuttle flight. This poor performance was a risk which management accepted. When pieces of foam are shaken off as the shuttle lifts off, no harm is done, but, on Columbia's last flight, a chunk of foam of about one and a half pounds broke off and hit the Space Shuttle Orbiter at over 400 miles per hour. This high speed collision was not discovered until a video recording; available the day after the launch, was played back, in slow motion, by engineers at NASA. In this video, foam hitting and jolting the Orbiter and disintegrating could be plainly seen 81.9 seconds after liftoff. Whether the Orbiter was damaged or not could not be determined from the video or from simulation experiments because the foam appeared to hit the underside of the wing or the wings leading edge, neither of which could be seen on the video.

The Orbiter successfully escaped the earth's atmosphere and reached orbital altitude. Columbia's scientific mission went forward as planned without event. Back on earth, engineers requested pictures of the underside of the Orbiter's wing from the U.S. Department of Defense's spy satellites. Since the accident was unplanned, there were no requirements to have spy satellites take pictures of the Orbiter. NASA

not independent or redundant systems as required. It was also discovered that the O-rings lost much of their capability to flex under cold conditions. In the hours before Challenger was scheduled to launch, on a nearly freezing day, engineers tried to stop the launch but were overruled by management. The launch went forward and, only about 47 seconds before the SRBs would have been jettisoned, the O-rings failed. Hot gasses burned through the O-rings of one of the SRBs and melted the SRB structure allowing a SRB to crash into the External Tank and ignite its fuel. This accident occurred about 73 seconds after Challenger's launch at an altitude of about 46,000 feet. The Challenger and her crew were lost.	Management refused to allow the engineering teams' request for pictures to be fulfilled on the basis that there was no requirement for pictures. As we now know, the chunk of foam that separated from the ET created a hole in the Orbiter's wing of sufficient size to destroy the Orbiter on reentry to the atmosphere. To make matters worse, there was, apparently, a viable possibility of rescuing the crew of Columbia by using the Space Shuttle Atlantis – if the pictures the engineers requested were taken and they showed the damage.

1- The Strategic Conception Phase	
The customers' interests are embodied in product and service requirements.	
Challenger – 1986	Columbia – 2003
"The O-rings of the Solid Rocket Boosters were not designed to erode. Erosion was a clue that something was wrong. Erosion was not something from which safety can be inferred."[143] (Richard Feynman, Appendix F of the Presidential Report) Morton Thiokol engineer, Roger Boisjoly, reportedly said "We all knew if the seals failed the shuttle would blow up."[144] NASA's performance specification "did not adequately take into account the	"3.3.3.2.8 ... the ET shall be designed to ensure the ET debris does not impact the Orbiter..."[146] "The assumption that only tiny pieces of debris would strike the Orbiter was also built into original design requirements, which specified that the Thermal Protection System (the tiles and Reinforced Carbon-Carbon, or RCC, panels) would be built to withstand impacts with a kinetic energy less than 0.006 foot-pounds. Such a small

---

[143] Report of the Presidential Commission on the Space Shuttle Challenger Accident, Appendix F, Washington, D.C., June 6, 1986. (Hereafter referred to as the Presidential Report)

[144] Boisjoly, Roger. 1986 Interview by Daniel Zwerdling of NPR. Downloaded 12/4/2013. http://www.npr.org/blogs/thetwo-way/2012/02/06/146490064/remembering-roger-boisjoly-he-tried-to-stop-shuttle-challenger-launch

known weather conditions that occur in Florida during the winter months"[145] (House Report, p. 50)	tolerance leaves the Orbiter vulnerable to strikes from birds, ice, launch pad debris, and pieces of foam." (CAIB, Vol. 1, p. 122)

2- The Recognition of Unacceptable Performance Phase
Unacceptable performance occurs and is observed.

Challenger – 1986	Columbia – 2003
"Neither Thiokol nor NASA expected the rubber O-rings sealing the joints to be touched by hot gases of motor ignition, much less to be partially burned. However, as tests and then flights confirmed damage to the sealing rings, the reaction by both NASA and Thiokol was to increase the amount of damage considered 'acceptable.'" (Presidential Report, Ch. 6, p. 121) NASA's Chief of the Solid Motor Branch, John "Miller signed and forwarded a second memo strenuously objecting to Thiokol's Solid Rocket Motor joint seal design. This memo, dated January 19, 1979, opened with: "We find the Thiokol position regarding design adequacy of the clevis joint to be completely unacceptable. . ." (Presidential Report, Ch. 6, p. 124) In a 1979 report of a trip to the Parker Seal Company, which manufactured the SRB's O-ring's, NASA representatives wrote "...their first thought was that the O-ring was being asked to perform beyond its intended design and that a different type of seal should he considered." (House Report, p. 52)	"Despite the design requirement that the External Tank shed no debris, and that the Orbiter not be subjected to any significant debris hits, Columbia sustained damage from debris strikes on its inaugural 1981 flight. More than 300 tiles had to be replaced." (CAIB, Vol. 1, p.122) "Over the life of the Space Shuttle Program, Orbiters have returned with an average of 143 divots in the upper and lower surfaces of the Thermal Protection System tiles, with 31 divots averaging over an inch in one dimension." (CAIB, Vol. 1, p. 122)

---

[146] Flight and Ground System Specification, NASA, Houston, TX. NSTS 07700, Vol. X - Book 1, Rev. M, 11/10/1998.

[145] Investigation of the Challenger Accident, House of Representatives, Report 99-1016, 10/29/1986. (Hereafter referred to as the House Report.

3- The Defect Repair Phase

Defects are repaired, when necessary, following use of a product or service. The process of repairing defective products or services improves to become a smooth operation.

Challenger – 1986	Columbia – 2003
The Solid Rocket Boosters were designed to be recovered from the ocean after each shuttle flight and reused. Rubber O-rings which are probably disposable – because they are a relatively inexpensive item – are replaced with new O-rings for the next shuttle flight.	Thermal tiles on the Space Shuttle, which were damaged, were replaced after each flight. The ET is non-reusable since it is designed to burn up in the atmosphere during each flight. New insulating foam is applied to a new ET for each flight. "most of the Shuttle Program's concern about Columbia's foam strike were not about the threat it might pose to the vehicle in orbit, but about the threat it might pose to the schedule."[147] (CAIB, Ch. 6, p. 139)

4- The Delusionary Phase

The fallacious beliefs that 1) fixing defects improves quality; 2) improving processes improves quality; or that 3) risks do not have a lifecycle, act to enable responsible parties to form a delusion with which they are psychologically and organizationally enabled to evade correction of product and service designs or conditions yielding poor or unacceptable performance.

Challenger – 1986	Columbia – 2003
The "Process Management Premise" fallacy[148] "The phenomenon of accepting for flight, seals that had shown erosion and blow-by in previous flights, is very clear. The Challenger flight is an excellent example. There are several references to flights that had gone before. The acceptance and success of these flights is taken as evidence of safety. But erosion and blow-by are not what the design expected. They are warnings that something is wrong. The equipment is not operating as expected, and therefore there is	Fixing defects improves quality fallacy "The first known bipod ramp foam loss occurred during STS-7, Challenger's second mission... a 19- by 12-inch piece of the left bipod ramp was missing... After the mission was completed, the Program Requirements Control Board cited the foam loss as an In-Flight Anomaly. Citing an event as an In-Flight Anomaly means that before the next launch, a specific NASA organization must resolve the problem or prove that it does not threaten the safety of the vehicle or crew.

---

[147] http://anon.nasa-global.speedera.net/anon.nasa-global/CAIB/CAIB_lowres_chapter6.pdf
[148] The Process Management Premise fallacy was introduced in chapter 1.

a danger that it can operate with even wider deviations in this unexpected and not thoroughly understood way." (Richard Feynman, Appendix F of the Presidential Report)

With this phenomenon, the worst case condition of the O-rings broadens the acceptable "in-family" historical performance range. People can now worry less about increasingly poor or unacceptable performance – as though the process improved due to worsening performance quality.

Comparing the organizational culture of one of NASA's organizations to another Mr. Feynman wrote: "... the computer software checking system and attitude is of the highest quality. There appears to be no **process of gradually fooling oneself while degrading standards so characteristic of the Solid Rocket Booster or Space Shuttle Main Engine safety systems.**" (Richard Feynman, Appendix F of the Presidential Report)

Content Coverage Error of Omission
Apparently, no effective process addressed the need to improve product design when it was discovered that performance was incapable of meeting the established design specifications.

"At the Flight Readiness Review for the next mission, Orbiter Project management reported that, **based on the completion of repairs to the Orbiter Thermal Protection System, the bipod ramp foam loss In-Flight Anomaly was resolved, or "closed."** However, although the closure documents detailed the repairs made to the Orbiter, **neither the Certificate of Flight Readiness documentation nor the Flight Readiness Review documentation referenced correcting the *cause* of the damage** – the shedding of foam [from the ET]." (CAIB, Vol. 1, p. 123)

Content Coverage Error of Omission
"...no process addresses the need to update a hazard analysis when anomalies occur." (CAIB, Vol. 1, p. 126)

As discussed in the USS Indianapolis case study, of chapter 2, the absence of a necessary procedure is, by definition, a failure in the Acceptance Theme of Performance. In this case, however, the omission is not simply an innocent oversight. Here, there is a significant philosophical flaw, or delusion, that "fixing defects improves quality" and that "risks do not evolve over time throughout their lifecycle" which would have prevented recognition that a procedure to refresh the hazard analysis was missing and recognition that the missing procedure was necessary. In the absence of delusional thinking, we would expect risk to influence decision makers to demand an updated hazard analysis. For example, when experts at the Parker Seal Company expressed the belief that their O-rings for the Solid Rocket Boosters were being expected to perform beyond their

	design "it was agreed that tests which more closely simulated actual conditions should be done."[149]

5- The Straw Man Phase

A history of lucky performance is set up as the basis for "Straw Man" requirements which are substituted for the customers' actual requirements. In this way, the customer's interests are disembodied from working requirements.

Challenger – 1986	Columbia – 2003
SRB contractor Morton "Thiokol and Marshall [NASA's George C. Marshall Space Flight Center] continued to rely on the redundancy of the secondary O-ring long after NASA had officially declared that the seal was a non-redundant single point failure." (Presidential Report, Ch. VI) With regard to erosion of O-rings, which is caused by "blow-by" of hot combustion gasses, SRB contractor Morton "Thiokol said in its conclusions that 'low temperature enhanced probability of blow-by [Shuttle flight] 51-C experienced worst case temperature change in Florida history.' Thiokol concluded that while the next Shuttle flight 'could exhibit same behavior,' nonetheless 'the condition is not desirable but is acceptable.'" (Presidential Report, Ch. VI)	"NASA has not followed its own rules and requirements on foam-shedding. Although the agency continuously worked on the foam-shedding problem, the debris impact requirements have not been met on any mission." (CAIB, Vol. 1, p. 130) "With each successful landing, it appears that NASA engineers and managers increasingly regarded the foam-shedding as inevitable, and as either unlikely to jeopardize safety or simply an acceptable risk. The distinction between foam loss and debris events also appears to have become blurred. NASA and contractor personnel came to view foam strikes not as a safety of flight issue, but rather a simple maintenance, or "turnaround" issue. In Flight Readiness Review documentation, Mission Management Team minutes, In-Flight Anomaly disposition reports, and elsewhere, what was originally considered a serious threat to the Orbiter came to be treated as "in-family," a reportable problem that was within the known experience base, was believed to be understood, and was not regarded as a safety-of-flight issue." (CAIB, Vol. 1, p. 122)

---

[149] Ray, William L. Memo, Visit to Precision Rubber Products Corporation and Parker Seal Company, 2/6/1979. Retrieved, on 5/30/2014, from http://history.nasa.gov/rogersrep/v5p1647.htm.

6- The Conformance to Straw Man Requirements Phase

Creativity and creative effort to correct poor or unacceptable performance is inhibited by commitment to the intention to conform to disembodied Straw Man requirements.

Challenger – 1986	Columbia – 2003
"The retention rationale [the basis for approval of waivers for critical items] is used to justify accepting the design "as is," in the Committee's view; its audits of the NASA review process discovered little emphasis on <u>creative</u> ways to eliminate potential failure modes."[150]	CAIB Member Douglas D. Osheroff is reported to have said "Foam appeared on lists of critical safety risks for every flight, but its defects were given a standing waiver; to the astonishment of many board members, such waivers were reviewed only annually, rather than during preparations for each flight."[151] **<u>Management by Fact</u>** "While rules and procedures were essential for coordination, they had an unintended but negative effect. Allegiance to hierarchy and procedure had replaced deference to NASA engineers' technical expertise." (CAIB, Vol. 1, p. 200)

7- The Acceptable Risk Phase

Poor or unacceptable performance, which has evaded correction, is masked as "acceptable risk" due to a growing history of luck and the growing history of defects proving to be repairable.

Challenger – 1986	Columbia – 2003
"NASA and Thiokol accepted escalating risk apparently because they "got away with it last time." As Commissioner Feynman observed, the decision making was: "a kind of Russian roulette. ... (The Shuttle) flies (with O-ring erosion) and nothing happens. Then it is suggested, therefore, that the risk is no longer so high for the next flights. We can lower our standards a little bit because we got away with it last time. ...You got away with it, but it shouldn't be done over and over again like that."	"F6.1-2 Foam-shedding, which had initially raised serious safety concerns, evolved into "in-family" or "no safety-of-flight" events or were deemed an "accepted risk." (CAIB, Vol. 1, p. 130) "...the pressure of maintaining the flight schedule created a management atmosphere that increasingly accepted less-than-specification performance of various components and systems, on the grounds that such deviations had not interfered with

---

[150] Aeronautics and Space Engineering Board, Post-Challenger Evaluation of Space Shuttle Risk Assessment and Management, National Academy Press, Jan. 1988, p. 2.

[151] Smith, R. Jeffrey, Mistakes of NASA toted up, Washington Post, Jul 13, 2003.

224

(Presidential Report, Ch. VI) "Thiokol included a report on damage incurred by the O-rings during [Shuttle] flight 51-C at the left forward and right center field joints. The right center joint had hot gas past the primary O-ring. Thiokol said that caused a concern that the gas seal could be lost, but its resolution was 'accept risk.'" (Presidential Report, Ch. VI) "NASA did not accept the judgment of its engineers that the design was unacceptable, and as the joint problems grew in number and severity NASA minimized them in management briefings and reports. Thiokol's stated position was that "the condition is not desirable but is acceptable." (Presidential Report, Ch. VI) "The notion of ACCEPTABLE EROSION was mentioned at a meeting of the Shuttle Projects Office Board for STS-41-C. Even though the joint was now classified as Criticality 1, which meant that failure of the joint could lead to the loss of the Shuttle and crew..." (House Report, p. 53)	the success of previous flights." (CAIB, Vol. 1, p. 24)

8- The Masking of Deviation as Variation Phase

To preserve the energy of a key strategy, significant deviations in performance, indicative of sophisticated risks with adverse trends, are often mischaracterized as extremes of variation or as "outliers" of a system of normal variation. In this way, the true nature of design or execution risks is masked, and causes of the risks remain, preserving the risk of poor performance.

Challenger – 1986	Columbia – 2003
"While flight anomalies involving the O-rings received considerable attention at Morton Thiokol and at Marshall, the significance of the developing trend went	"According to the STS-113 Flight Readiness Review, foam loss was not considered a safety-of-flight issue, but rather an "accepted risk," which meant that the

225

unnoticed." (Presidential Report, Ch. VII)

"NASA elected to regard a certain degree of erosion or blow-by as "acceptable." To make matters worse, confidence was mistakenly obtained from a mathematical model which suggested that if the erosion did not exceed a specific depth, the O-ring would still seal that joint. In cases where the erosion did exceed the maximum predicted by the model, NASA expanded its experience base to cover this increased damage." (House Report, pp. 62-63.)

"The Challenger commission determined that before the accident, for example, NASA did not understand the technical basis of its "O-ring" troubles[152] and failed to detect that they were getting worse. In retrospect, the commission said, "this trend is easily recognizable." (Mistakes Of NASA toted up)

"A careful analysis of the flight history of O-ring performance would have revealed the correlation of O-ring damage and low temperature. Neither NASA nor Thiokol carried out such an analysis; consequently, they were unprepared to properly evaluate the risks of launching the 51-L mission in conditions more extreme than they had encountered before." (Presidential Report, p. 149)

threat was not zero but rather a known and acceptable risk."[153]

"F6.1-10 NASA failed to adequately perform trend analysis on foam losses. This greatly hampered the agency's ability to make informed decisions about foam losses." (CAIB, Vol. 1, p. 131.)

---

[152] This use of the word "troubles" is interesting because, as we now understand risk, the word risk would be more appropriate here. It's possible that the authors didn't associate the word risk with something that was "getting worse" since risk is traditionally thought of in static terms.

[153] Columbia's Last Mission, Academy of Program/Project & Engineering Leadership, NASA. Downloaded on 6/29/2014 from http://appel.nasa.gov/knowledge-sharing/publications/columbias_last_mission-html/ p.10.

9- The Fearlessness Phase

Unacceptable performance, bearing risk, masquerades into the whirlwind of acceptable performance, gravitating towards a common strategy, and creates the false impression of energizing the strategy, while it is, instead, harboring a potential for unrestrained adversity.

Challenger – 1986	Columbia – 2003
In July 1985, Roger Boisjoly, a Thiokol engineer, wrote to his boss, Mr. Bob Lund, V.P. Engineering at Thiokol "The mistakenly accepted position on the joint problem was to fly without fear of failure... This position is now changed as a result of the [51-B] nozzle joint erosion which eroded a secondary O-ring with the primary O-ring never sealing. If the same scenario should occur in a field joint (and it could), then it is a jump ball [Probability ≈ 50%] whether as to the success or failure of the joint because the secondary O-ring cannot respond to the clevis opening rate and may not be capable of pressurization. The result would be a catastrophe of the highest order-loss of human life." (Roger Boisjoly and J. R. Kapp, Morton Thiokol Interoffice memo to R.K. Lund V.P. Engineering, July 31, 1985, Presidential Report , Vol. 4) "...the management of NASA exaggerates the reliability of its product, to the point of fantasy." (Richard Feynman, Appendix F of the Presidential Report)	"It seems that Shuttle managers had become conditioned over time to not regard foam loss as a safety-of-flight concern." (p.125, Columbia AIR, vol1, Aug. 2003)

10- The Black Operations to Liberate Black Swans Phase[154]

Management reflexively, and irresistibly, acts to suppress risk warnings in order to defend and sustain a key strategy. The compulsion to avoid unmasking deviation and decisions which created straw man requirements becomes more controlling over behavior than the legitimate responsibility to respond to risk and prevent adversity. In this way, risks with the potential for rare and dangerous "Black Swan Events" are liberated to evolve under inadequate control.

---

[154] The phrase "Black Swan event" is used by statisticians to describe an unlikely event.

Challenger – 1986	Columbia – 2003
The Risk Warning	**The Risk Warning**
In Roger Boisjoly's July 1985 memo to his boss, Mr. Bob Lund, he wrote "It is my honest and very real fear that if we do not take immediate action to dedicate a team to solve the problem, with the field joint having the number one priority, then we stand in jeopardy of losing a flight along with all the launch pad facilities."[155]	The day after Columbia's launch, engineers at the Kennedy Space Center, reviewing high resolution video, "noticed a debris strike at 81.9 seconds after launch. A large object from the left bipod area of the External Tank struck the Orbiter, apparently impacting the underside of the left wing near RCC panels 5 through 9. The object's large size and the apparent momentum transfer concerned Intercenter Photo Working Group personnel... This concern led the Intercenter Photo Working Group Chair [Bob Page] to request, in anticipation of analysts' needs, that a high-resolution image of the Orbiter on-orbit be obtained by the Department of Defense [the Air Force]." (CAIB Vol. 1, p.37) Perhaps due to urgency, the preferred process for making this request wasn't followed – but the back-channel request for imagery was working.
Decisions Managers Want to Keep Masked	
"Since December, 1982, the O-rings had been designated a "Criticality 1" feature of the Solid Rocket Booster design, a term denoting a failure point-without back-up-that could cause a loss of life or vehicle if the component fails. In July 1985, after a nozzle joint on STS 51-B showed erosion of a secondary O-ring, indicating that the primary seal failed, a launch constraint was placed on flight 51-F and subsequent launches. These constraints had been imposed and regularly waived by the Solid Rocket Booster Project Manager at Marshall [NASA's Marshall Space Flight Center], Lawrence B. Mulloy."	
	In support of his requests for imagery, the shuttle's chief structural engineer, Rodney Rocha, e-mailed his boss "Without better images it will be very difficult to even bound the problem and initialize thermal, trajectory, and structural analyses. Their [the Debris Assessment Teams] answers may have a wide spread ranging from acceptable to not--acceptable to horrible, and no way to reduce uncertainty."[156]
In a meeting between NASA and Thiokol, on January 28, 1986, the evening prior to Challenger's launch, Thiokol engineers presented a series of charts leading to their recommendation not to launch due to the freezing temperatures which had dipped as low as 18 degrees Fahrenheit. NASA challenged some of the data presented by engineers on their charts. As Mr. Boisjoly testified: "I was asked to quantify my concerns, and I said I couldn't. I	**Suppression**
	"On the seventh day of the flight, January 22, just as the Air Force began to

[155] Roger Boisjoly and J. R. Kapp, Morton Thiokol Interoffice memo to R.K. Lund V.P. Engineering, July 31, 1985, Presidential Report , Vol. 4
[156] Rocha, Alan. "STS-107 Wing Debris Impact, Request for Outside Photo-Imaging Help." E-mail from Alan Rocha to Paul Shack and other NASA employees, 1/21/2003. Presented in CAIB, Vol. 1, p. 151.

couldn't quantify it. I had no data to quantify it, but I did say I knew that it was away from goodness in the current data base." (Presidential Report, Ch. V) The decisive recommendation of the engineering team at Thiokol was that the "O-ring temp must be ≥ 53 °F at launch" (Initial Thiokol recommendation Chart presented by Mr. Lund, Presidential Report, Ch. V) **Suppression** When told of the Thiokol engineering team's recommendation not to launch, due to the unprecedented cold, Mr. Lund testified that "Mr. Hardy [George Hardy, NASA's Deputy Director of Science and Engineering] said he was appalled that we would make such a recommendation." (Presidential Report, Ch. V) Mr. Mulloy, who was well acquainted with the seriousness of the O-ring risk, since it was he who repeatedly issued waivers accepting their poor performance, said: "My God, Thiokol, when do you want me to launch, next April?" (Presidential Report, Ch. V) Thiokol managers and two engineers broke out of the meeting with NASA and the engineering team for a "caucus" to discuss the launch decision further. Roger Boisjoly testified that "After Arnie [Thiokol Engineer, Arnold Thompson] and I had our last say, Mr. Mason [Jerry Mason, Thiokol Senior Vice President] said we have to make a management decision. He turned to Bob Lund and asked him to take off his engineering hat and put on his management	move on the Kennedy engineers' back-channel request for photographs, [Mission Management Team Chair] Linda Ham heard to her surprise that this approach (which according to front-channel procedures would have required her approval) had been made. She immediately telephoned other high-level managers in Houston [Texas] to see if any of them wanted to issue a formal "requirement" for imagery, and when they informed her that they did not, rather than exploring the question with the Kennedy engineers [in Florida] she simply terminated their request with the Department of Defense. This appears to have been a purely bureaucratic reaction."[157] "Linda Ham had several reasons for canceling the request for imagery. She had confirmed that no one had a "requirement" for using DoD assets to obtain on-orbit imagery." "Linda Ham said it [imagery] was no longer being pursued since even if we saw something, we couldn't do anything about it. The Program didn't want to spend the resources." (CAIB, p. 154) **Decisions Managers Want to Keep Masked** "Having at hand a previously accepted rationale – reached just one mission ago – that foam strikes are not a safety-of-flight issue provides a strong incentive for Mission managers and working engineers to use that same judgment for STS-107 [the Columbia flight]. If managers and engineers were to argue that foam strikes are a safety-of-flight issue, they would contradict

[157] Langewiesche, William. Columbia's Last Flight: The inside story of the investigation—and the catastrophe it laid bare. The Atlantic Monthly November 2003.
http://www.theatlantic.com/past/docs/issues/2003/11/langewiesche.htm

Challenger – 1986	Columbia – 2003
hat. From this point on, management formulated the points to base their decision on. There was never one comment in favor, as I have said, of launching by any engineer or other non-management person in the room before or after the caucus. ...I left the room feeling badly defeated, but I felt I really did all I could to stop the launch. I felt personally that management was under a lot of pressure to launch and that they made a very tough decision, but I didn't agree with it. One of my colleagues that was in the meeting summed it up best. This was a meeting where the determination was to launch, and it was up to us to prove beyond a shadow of a doubt that it was not safe to do so." (Presidential Report, p. 94)	an established consensus that was a product of the Shuttle Program's most rigorous review – a review in which many of them were active participants." (CAIB, Vol. 1, p. 150)

<table>
<tr><td>

Black Ops

The chairman of the Presidential Commission asked Mr. Lund "How do you explain the fact that you seemed to change your mind when you changed your hat? To which Mr. Lund replied "I guess we have got to go back a little further in the conversation than that. We have dealt with Marshall [NASA] for a long time and have always been in the position of defending our position to make sure that we were ready to fly, and I guess I didn't realize until after that meeting and after several days that we had absolutely changed our position from what we had been before... We had to prove to them that we weren't ready, and so we got ourselves in the thought process that we were trying to find some way to prove to them it wouldn't work, and we were unable to do that. We couldn't prove absolutely that that motor wouldn't work." (Presidential Report, p. 95)

</td><td>

Black Ops

"Analysts on the Debris Assessment Team were in the unenviable position of needing images to more accurately assess damage while simultaneously needing to prove to Program managers, as a result of their assessment, that there was a need for images in the first place." (CAIB, Vol. 1, p. 157)

Requests for imagery were a response to the realization that a large chunk of foam debris, possibly weighted down with ice, struck the orbiter at between 400 and 550 miles per hour. Of course, no one had a requirement for this accident to occur and, therefore, there was no requirement to ask the DoD to take pictures.

"The engineers found themselves in the unusual position of having to prove that the situation was *unsafe* - a reversal of the usual requirement to prove that a situation is *safe*." (CAIB, Vol. 1, p. 169)

"When [the shuttle's chief structural engineer Rodney] Rocha heard about their decision [to not request photos], he angrily confronted [his boss, head of the shuttle integration office, Paul] Shack and told him he was ignoring an obvious risk, sources said. Shack responded by calling him an alarmist."[158]

</td></tr>
</table>

---

[158] Smith, R. Jeffrey, Mistakes of NASA toted up, Washington Post, Jul 13, 2003.

Challenger - 1986	Columbia - 2003
The Unmasking "The Commission concluded that the Thiokol Management reversed its position and recommended the launch of 51-L, at the urging of Marshall [NASA] and contrary to the views of its engineers in order to accommodate a major customer." (Presidential Report, Ch. 5)	"... the astronauts had been told of the [foam] strike, but almost as if they were children who didn't need to be involved in the grown-up conversation."[159] If the astronauts had been told the analysis was incomplete and that safety of flight was inconclusive they may have asked the troublesome question – "Has anyone taken a good photograph of the wing?" **The Unmasking** "We are convinced that the management practices overseeing the Space Shuttle Program were as much a cause of the accident as the foam that struck the left wing." (CAIB, Vol. 1, p. 11)

11- The Concluding Phase
The project completes successfully or an accident occurs.

Challenger - 1986	Columbia - 2003
A catastrophic accident occurred.	A catastrophic accident occurred.

---

[159] Langewiesche, William. Columbia's Last Flight

Booster Rockets Challenger – 1986	External Tank Columbia – 2003

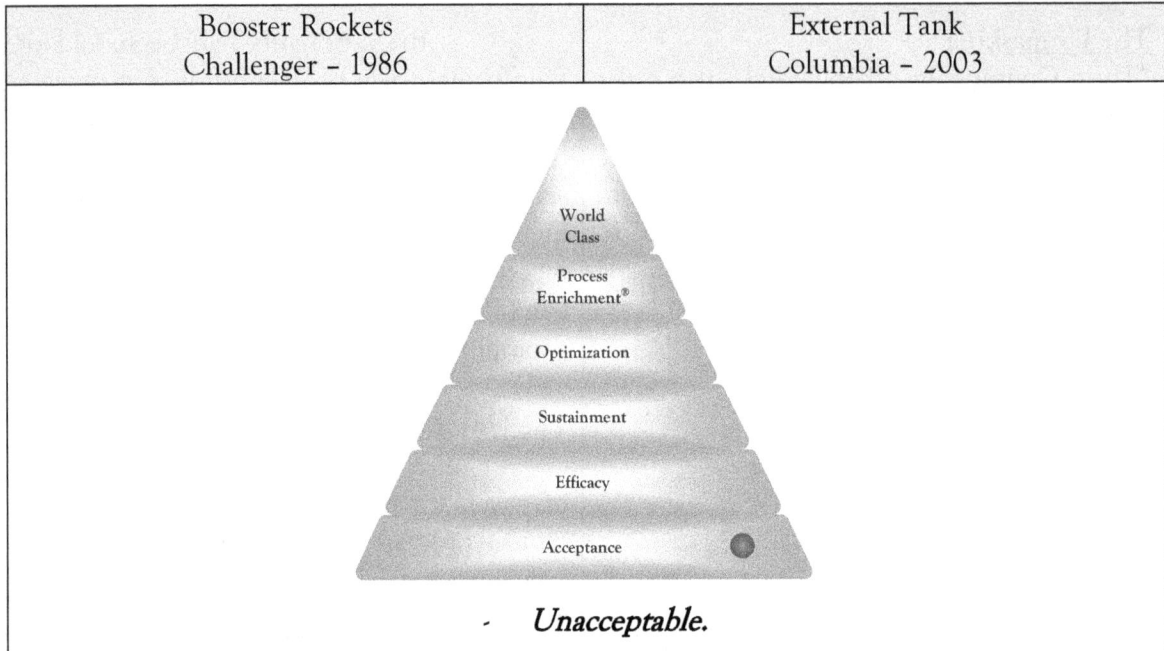

World
Class

Process
Enrichment®

Optimization

Sustainment

Efficacy

Acceptance

- *Unacceptable.*

In both of the Space Shuttle accidents, in theory, tragedy could have been avoided if the Acceptance Theme question – "Is it acceptable?" – was answered in the correct way and, absent the Fearlessness Effect, an attempt was made to circumvent, counteract, and triumph over risk by restoring, sustaining, and maximizing quality in response to the potential for adversity. We can see that this realization could have been reached at the times of these accidents just as well as, afterwards, in hindsight. By allowing our creative effort to retain correctable design flaws, which are sensible to correct, and by the misguided use of procedures as weapons to suppress warnings and evidence of poor performance, we may find, especially on safety-critical applications that our strategy may fail with tragic results.

A valuable lesson from NASA's engineering managers, learned during construction of the Lunar Module for the Apollo 11 mission, expresses the criticality of understanding the nature of risk and having creative vision beyond conformance to requirements.

"We were getting ready to go to the moon on Apollo 11. The inertial measurement unit on the lunar module was perfect, no drift rate. All of a sudden it started drifting high but not out of spec. We, the Draper Labs or the MIT Instrumentation Lab and the subsystem managers, all went to George Low [Manager of the Apollo Spacecraft Program Office] and told him he did not have to change the IMU out on the lunar module. Very risky. The lunar module was made out of Reynolds wrap almost. And George Low looked at us. He said, "You may be right, but I'm going to change it out." It was telling a message. It was telling a message that it was drifting ~ not out of spec but it started doing something different. I'll remember that as long as I live as a thing that you need to think about."[160]

Aaron Cohen
Manager of Apollo Modules

---

[160] CAIB Vol. VI, pp. 239-240

Chapter 5:

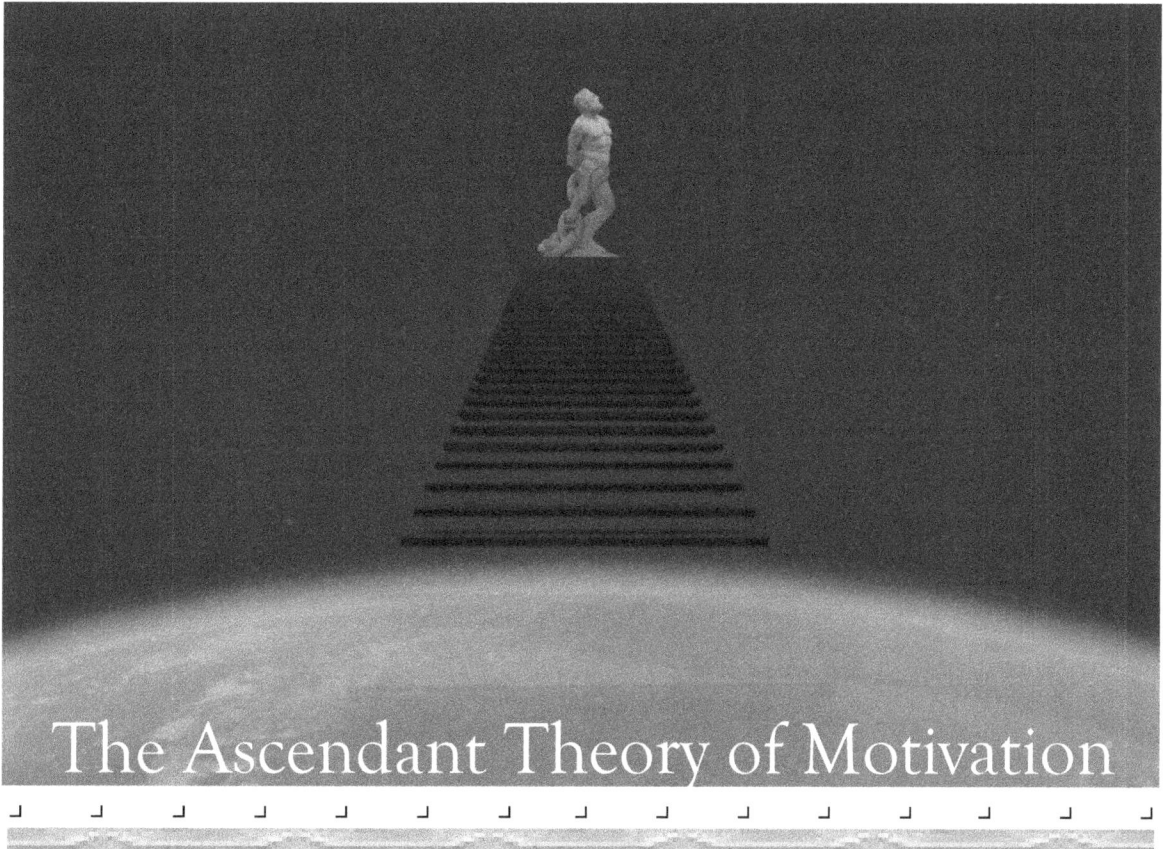

The Ascendant Theory of Motivation

The most common, elementary, and, perhaps,
defining property of living beings
is that we each possess a fundamental need
to be on the quest for ascendant quality.

Chapter 5:

The Ascendant Theory of Motivation

PERSPECTIVES

▶ The fundamental and basic needs that incentivize human motivation

▶ Life's Universal Incentive

▶ The three principle tenets of the Ascendant Theory of Motivation

▶ The impact of motivation on creative effort

▶ The Ascendant Theory of Motivation and the kindred theories and studies of human motivation

    o Maslow's Hierarchy of Needs theory

        ▪ Exception 7 to Maslow's hierarchy of needs theory

    o Herzberg's Two-Factor theory

▶ Satisfaction, dissatisfaction, motivation, personal strategy and Coascendancy in the Ascendant Theory of Motivation

▶ The Coascendancy Totem illustrating how productivity, employee satisfaction, and customer value are improved by conditions which attract group behavior towards organizational strategy.

It is the ageless testimony of humankind's ever-evolving creative effort that stands as the best evidence that we are a people deeply devoted to the quest for ascendant quality. Our need to be on the quest for ascendant quality is an innate gift which serves as the most treasured incentive for our motivation to evolve and to create, with ever-improving quality and diversity, all that we desire for an improved quality of life. It is because the need to be on the quest for ascendant quality captures the essence of our nature so completely that, in the Process Enrichment philosophy's Ascendant Theory of Motivation, we call the need to be on the quest for ascendant quality a fundamental need. As a fundamental need, the need to be on the quest for ascendant quality serves as the predominant incentive for our motivated behavior – having more influence over human motivation, intentions, and behavior than any other need. This belief is expressed as the first of three principal tenets of the Ascendant Theory of Motivation.

LIFE'S UNIVERSAL INCENTIVE

Living beings possess a fundamental need to be on the quest for ascendant quality.

The relationship of motivation to quality is central to quality's Process Enrichment Theme of Performance. In quality's Process Enrichment Theme of Performance, it is critical that all parties to a Coascendant relationship sustain motivation not only towards the incentives they pursue independently, but also in the performance of the processes used to attain the common incentives of their common venture. This is best achieved, for example, in commerce, when customers have suppliers that are "easy to do business with," as the expression goes, who make it easy for customers to work with the supplier's organization, and also with the products and services provided, in order to attain the one best way of performing their strategies. For suppliers, quality in the Process Enrichment Theme requires their employees to be motivated towards performance of their creative effort through an efficient organizational system that makes creative effort possible, efficient, and personally rewarding. As a consequence of Life's Universal Incentive we find that conditions serving to create success and progress in our quest for ascendant quality provide the experience of gratification which further reinforces and motivates the desire to recreate conditions of success and progress, free of the risk of poor performance. As we come to understand the fundamental and basic needs which incentivize and energize our aspirations we may gain the wisdom, the will, and the way to approach bringing the rewards of satisfaction and gratification to ourselves and to our creative partners with whom we seek to build Coascendant relationships.

We may never be able to establish, by unassailable facts, the causes or the combination of causes that motivate living beings, however, as in many scientific endeavors, we exercise the talent of our minds to weigh the information we are confident in and stand it up as a theory from which we can enjoy a benefit – until someone who interprets the information differently builds upon it or proposes an alternative theory. This is a beneficial way for innovative thought to progress in a science, such as psychology, where it is customary to present well founded beliefs as theories rather than as facts.

The belief underlying one of psychology's most enduring theories of motivation was expressed in a famous quote of Dr. Abraham Maslow, one of the last century's most renowned psychologists.

"Man is a perpetually wanting animal."[161]

Abraham Maslow

In the interest of science, the Ascendant Theory of Motivation enhances Dr. Maslow's observation by theorizing that man is, indeed, a perpetually wanting animal and that *what man perpetually wants is ascendant quality*. This new theory provides a rational strategic purpose for our "wants," which we can easily accept as valid, because, as we know, even as we attain quality excellence, in every field of endeavor, we want to perpetually ascend from our accomplishments to ever higher plateaus of quality – surpassing the limits of quality excellence that previously restrained us – for the experience of gratification, for the temporary sense of relief and freedom from pursuit of our perpetual, fundamental need, and, also, for the improvement in our quality of life that we expect to attain.

---

[161] Abraham Maslow, A Theory of Human Motivation, July 1943, Psychological Review, Vol. 50, 370-396.

The irrepressible nature of our needs, which drive our imagination and animate our desires to surmount our limitations – even including the limitations imposed by our lifespan – was epitomized in the novel Frankenstein, written by Mary Shelley in 1818. In this classic and familiar story, in pursuit of ascendant quality, Dr. Frankenstein succeeded in creating a living creature from the dead. Success in this extraordinary endeavor brought the unintended and horrific consequences which created the angst in the character of Dr. Frankenstein. In a soliloquy lamenting mans' burden of possessing superior "sensibilities," and capabilities for reasoned thought, through Dr. Frankenstein's voice, Ms. Shelley expresses the nature of the challenges, triumphs, and sorrows accompanying mankind on the journey along the ever evolving path of the quest for ascendant quality.

"Alas! why does man boast of sensibilities superior to those apparent in the brute; it only renders them more necessary beings. If our impulses were confined to hunger, thirst, and desire, we might be nearly free; but now we are moved by every wind that blows, and a chance word or scene that that word may convey to us.

We rest. A dream has power to poison sleep;

We rise. One wandering thought pollutes the day;

We feel, conceive or reason, laugh or weep;

Embrace fond foe, or cast our cares away:

It is the same! For, be it joy or sorrow,

The path of its departure still is free:

Man's yesterday may ne'er be like his morrow;

Nought may endure but Mutability."[162]

Dr. Frankenstein
quoting from the poem Mutability by Percy Shelley
in the novel Frankenstein by Mary Shelley
1818

---

[162] Mary Shelley, Frankenstein; or, the Modern Prometheus, 1818 (London, England: Lackington, Hughes, Harding, Mavor & Jones)

In this excerpt from the novel Frankenstein, we find Dr. Frankenstein weighing the value of man's sensibilities and capabilities for reasoned thought. It was because of Dr. Frankenstein's sensibilities and talents as a doctor that he was able to create a creature that transcended death. Unintentionally, however, by this point in the story, through its violent acts, his creation had become a monster. Under these circumstances, Dr. Frankenstein's sensibilities and reasoning brought him anguish, instead of gratification, in consideration of the consequences of what his sensibilities and talents enabled him to create. Dr. Frankenstein's thoughts turn to a poem called "Mutability" in which he finds solace in reflecting on a truism – nothing endures except change – and he recognized that the path of change is continuous and so he found the resilience to pursue a solution to the dilemma created by his monster.

Through Dr. Frankenstein's soliloquy, we see that, paradoxically, like the constancy of change, as much as it is in our nature to seek freedom, we can never be free of our fundamental nature. As a fundamental part of our nature, our need to be on the quest for ascendant quality is a persistent part of who we are, and it is the incentive which inspires us and brings our motivations to life. As Ms. Shelley's Dr. Frankenstein revealed, accepting risks and applying our talents in order to transcend the quality excellence that restrains us is an impulse of our sensibilities at a higher level than the impulses of "hunger, thirst, and desire" that animate the "brute" creatures of the earth. This insight is informative about the concept of motivation and it is also of historical interest because it shows us that, nearly 200 years ago, Ms. Shelley outlined an understanding of a hierarchy of impulses which is very similar to the Hierarchy of Needs theory developed through the research of Abraham Maslow in the 1940's – which stands today as, perhaps, the most prominent theory of human motivation.

The evidence of our desire to transcend the limits imposed upon us by quality excellence is abundant not only in fiction, but in our real-life. We have chanced dozens of forms of government – each of which has boasted of the potential for improved freedoms; we speculate on the tools of technology which boast of the ability to free us from our chores; we speculate on medical research – which boasts of the potential for longer life; and, closer to our hearts, we use cleverness in our courting ritual – which boasts of the potential for establishing a Coascendant relationship. Consider, for example, the Coascendant quality of the

240

following Valentine's Day message I posted for my girlfriend, thirty years ago, in a local newspaper.

"If more we are together,
less can we be apart."

This was, perhaps, my most successful use of strategic communication to build Coascendancy. As we've seen in our life experiences, and as we've seen with Mary Shelley using a poem from her husband, Percy Shelley, in her book, Coascendant relationships can sometimes lead to a better quality of life and can occasionally lead to significant creative achievements. Although the pursuit of ascendant quality has not enabled us to transcend death, as in the novel Frankenstein, scientists are hard at work and are meeting with success in wide ranging efforts to prolong life. The ability of technology to improve civilizations throughout the world is growing at a great pace. We have also made great strides in forming civilizations that join in the Coascendant purposes of enhancing our freedoms, capabilities, commerce, and quality of life – within reasonable constraints.

If we look at how far science and civilization have come in the last few thousand years, due to the quest for ascendant quality, we may imagine that, in the biology of species, which evolve over thousands of thousands of years, the presence of the fundamental need to be on the quest for ascendant quality may also have a profound effect. One way that the need to be on the quest for ascendant quality may affect the biology of species is through strategic mate selection. When members of a species understand what ascendant quality means to them, they may opportunistically select mates that they perceive to present these strategically desirable qualities. In this case, when desirable variations or deviations occur in a species, these features or capabilities may become the attractive innovations or ascendant quality through which differing living beings become more or less attractive to potential mates. In this way, the desirable and attractive variations or deviations of a species are propagated to offspring in a continuum of controlled change.[163] One might theorize that evolution of a

---

[163] Modern science currently explains the changes in the gene patterns of species as a series of gene mutations, however, if our genes mutate for no ascendant, controlled, purpose, it's likely that our DNA would eventually be damaged beyond the capability to sustain life. Charles

species occurs purposefully and strategically – in much the same way that the standard of quality excellence for a product or service evolves through ascendant quality over time. Both forms of evolution depend on the capabilities to recognize ascendant quality and to transform, to the extent possible, beyond the current level of quality excellence, given the talents, resources and conditions available. While the true causes of evolution of a species will probably remain forever debatable, we can readily appreciate that quality in creative effort is evolving all around us because we have all witnessed and contributed to this evolution in our own way.

As a consequence of our need to be on the quest for ascendant quality, which is an ever-present influence on our world-view, we seek the gratification that comes from satisfaction of this need through various purposeful and exploratory missions – each producing varying degrees of gratification. Since our attainment of gratification requires these missions, we can infer that it is in our nature to be of a purposeful, continuously learning, adaptive and creative character – a character that readily contrives and executes quality maximizing and risk minimizing strategies to achieve our aims. It is with these traits that we can best achieve success in our missions and attain the gratification we desire. As we compose strategies which best fulfill our missions, and efficiently and reliably attain the gratification we desire, our worldview evolves, and our baseline of quality excellence, which is the standard upon which we gage further advancements, also evolves.

Revisiting Maslow's Hierarchy of Needs Theory

Dr. Abraham Maslow's Hierarchy of Needs theory is probably the most prominent modern-day theory of human motivation. The acceptance by the academic community of Dr. Maslow's theory, since its publication in 1943, is attested to by the fact that a large percentage of all American college graduates, over the last few decades, have been exposed to Maslow's Hierarchy of Needs theory at some point in their education. Dr. Maslow's Hierarchy of Needs theory

---

Darwin's theory of natural selection hypothesizes that evolution occurs through "survival of the fittest," however, if this were true, we might all be creatures that have shells like turtles', teeth like sharks', and we might emit deadly nerve gas like a walking octopus.

242

provides insight to the distinct types of needs – which he calls basic needs – as well as to their hierarchical order.

The Ascendant Theory of Motivation transforms and enriches Dr. Maslow's theory by introducing new insight to human motivation. A key insight is that by recognizing the quest for ascendant quality as a fundamental need we now have a theory that can explain the incentive which motivates progression through each basic need, and upwards through the Hierarchy of Needs, to motivation at the higher level needs. Our newly defined concept of strategy is also useful because where we find people pursuing incentives we find strategies and, since we now know the elements of strategy, this knowledge will enable us to better explain motivated behavior. Looking well beyond the intentions of the research presented in this book, there may be potential to use the knowledge of motivation and strategy that has been introduced here as diagnostic tools to compare rational and rationally deviant strategies in order to diagnose conditions of mental health. This usage of the concept of strategy would not be unexpected since the field of psychiatry often uses a variety of checklists and analysis to diagnose conditions of mental health.

Following a brief explanation of Maslow's Hierarchy of Needs, in his own words, excerpted from a 1943 paper introducing his theory, we discuss enhancing our ability to understand human motives and behaviors in an increased number of circumstances and in useful ways.

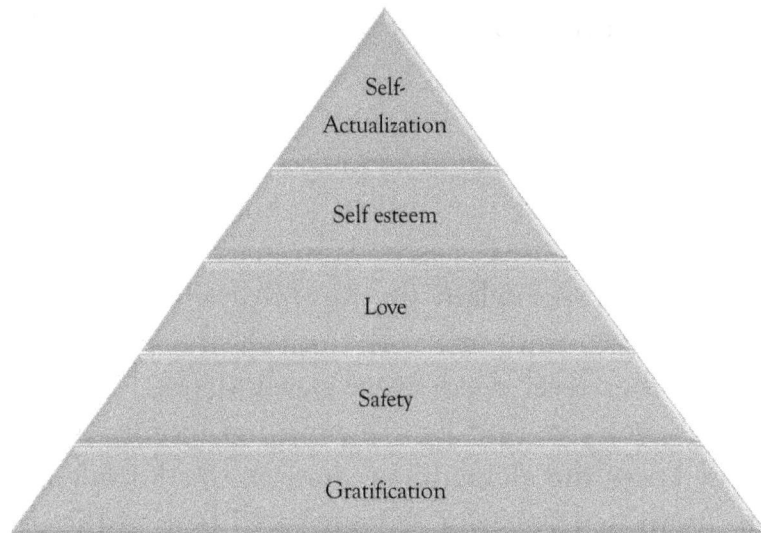

+ Conditions that make satisfaction of these needs possible

Figure 7-1, Maslow's Hierarchy of Needs, as originally presented by Dr. Abraham Maslow in 1943.

"There is a hierarchy of five sets of goals or purposes or needs which are set in the following order of prepotency. [164]

- First, satisfaction or gratification of body needs; the basic physiological and sensual gratifications.
- Second, the safety needs—insurance against pain, against danger to life, against jeopardy of bodily integrity, against overwhelming threats. There is usually a further component in this desire for safety in that it implies a world that can be counted on, that is familiar, and that is manageable. In a word, a world that does not threaten and that is not dangerous.
- Third, love, affection, warmth, acceptance, a place in the group.
- Fourth, desire for self-esteem, self-respect, self-confidence, for the feeling of strength or adequacy. Since this is partially a product of the respect of other people we may add also as partial goals prestige, respect from other people, etc.

---

[164] (Note: Dr. Maslow appears to use the words "goals", "purposes", "needs", and "wants" interchangeably.)

- Fifth, self-actualization, self-fulfillment, self-expression, working out of one's own fundamental personality, the fulfillment of its potentialities, the use of its capacities, the tendency to be the most that one is capable of being.

In addition, the individual will tend to want and to strive for all the conditions which make these satisfactions possible, e.g., freedom, full information, justice, order, etc.

We can describe these briefly by speaking of the fundamental goals of gratification, safety, love, self-esteem, and self-actualization. The first is prepotent over the second, which is prepotent over the third, which is prepotent over the fourth, which is prepotent over the fifth. This statement of prepotency is a statement of average findings in a large sampling of our population. This prepotency means primarily that where all the wants are unsatisfied, the one that will tend to dominate consciousness and behavior will be the search for gratification. It is often found further that a person with all the wants unsatisfied is actually not aware of having any wants other than the physiological wants. He has only the main conscious goals of getting food, water, and most often makes the naive assumption that if he is satisfied in these wants he will then be happy and contented. But this is not so. When the most dominant want is satisfied, a new one emerges into consciousness, usually the next higher in the hierarchy of prepotency. If this second want is satisfied then a third want appears from nowhere. Ordinarily the satisfaction of these wants are not altogether mutually exclusive but are so only as a matter of degree. The average member of our society is probably partially satisfied and partially dissatisfied in all of his wants."[165]

Dr. Abraham Maslow
1943

---

[165] MASLOW, A. H., A preface to motivation theory. Psychosomatic Med., 1943, 5, 85-92.

The first of three significant philosophical departures from Dr. Maslow's theory is the proposition that the need to pursue the Quest for Ascendant Quality is a more fundamental need then the basic physiological, safety, love, self-esteem and self-actualization needs Dr. Maslow identified. In other words, it may be more correct to say that the needs in Dr. Maslow's Hierarchy of Needs are basic needs through which the fundamental need, to pursue the quest for ascendant quality, attains a measure of satisfaction – bringing the feeling of gratification. Proof of this proposition can be found by reassessing and re-interpreting unexplained exceptions that Dr. Maslow identified in his theory. If Life's Universal Incentive explains one or more of these exceptions, without degrading other aspects of Maslow's Theory, then we will have successfully improved upon Dr. Maslow's theory.

Dr. Maslow composed the ordering of the Hierarchy of Needs because, based on his research, he believed this order represented the true order of needs. Dr. Maslow wrote: "This statement of prepotency is a statement of average findings in a large sampling." Dr. Maslow recognized, within his sampling, 7 exceptions to the ordering of needs in his Hierarchy of Needs. Dr. Maslow described the 7th such exception as follows:

> "... Perhaps more important than all these exceptions are the ones that involve ideals, high social standards, high values and the like. With such values people become martyrs; they give up everything for the sake of a particular ideal, or value..."[166]

Dr. Maslow
1943

This 7th exception appears to describe the fundamental need that we have been calling "the need to pursue the quest for ascendant quality" quite accurately. Dr. Maslow observes people "giving up" their basic needs in pursuit of it – this is exactly what we would expect to find if, indeed, our theory of a more fundamental need is correct. To determine whether or not the need to pursue the quest for ascendant quality is actually a more fundamental need than the

---

[166] Abraham Maslow, A Theory of Human Motivation, July 1943, Psychological Review, Vol. 50, 370-396

basic needs Dr. Maslow identified, let's examine them one at a time to see if circumstances and behaviors we're familiar with make what Dr. Maslow called exceptions to his theory look unexceptional in assumption of the presence of a more fundamental need to pursue the quest for ascendant quality.

A need more compelling than the Physiological Needs

Under the right circumstances, a person focused on gratification of physiological needs may choose to give up or suspend their physiological needs for pursuit of ascendant quality. For example, consider the acts of kamikaze pilots, suicide bombers, protesters who perform self-immolation, political prisoners on hunger strikes, etc. These self-sacrificing acts are known to be performed for an individual's need to serve what they and their sympathizers believe to be an ascendant purpose.

A need more compelling than the Safety Needs

A safety-focused person may suspend their safety need for pursuit of ascendant quality. For example, firefighters are likely to put their own safety at risk by running into fiery homes to perform the ascendant act of saving another life. Police, military, and doctors treating contagious patients exhibit similar ascendant behaviors. Another example is found with extremely elderly people with life threatening conditions who would rather live at home, at high risk, than live in a nursing home.

A need more compelling than the Love Needs

A love-focused person may suspend their love needs for pursuit of ascendant quality. For example, sometimes we can find, in a divorce, where both parents love the children and need the children's love, that the parent less able to take care of their children gives custody to the more capable parent. As another example, employees often get better jobs and quit their old jobs even when they work with or for friends and family. Even more telling is how unusual it is for someone to quit a job to work *with* friends or family – perhaps due to the more compelling need to be free to grow and ascend.

A need more compelling than the Self-esteem Needs

An esteem-focused person may suspend their esteem-need for pursuit of ascendant quality. For example, we may find that even if our company gives us a promotion that satisfies our need for self-esteem, if they've given us the promotion and require us to act unethically, we will quit the job. As another example, sometimes the way-too-young marry the way-too-old and rich – despite community ridicule – in order to attain the improved quality of life their rich spouses' money can buy.

A need more compelling than the Self-actualization Needs

An actualization-focused person may suspend their actualization need for pursuit of ascendant quality. For example, George Washington refused to be king of the U.S. and he also refused a third term in office. As another example, sometimes we will read the news and find that a famous actor, one who has presumably achieved a high degree of self-actualization, has gone to an overseas trouble spot to negotiate the release of hostages.

As this analysis indicates, a reasonable theory explaining Dr. Maslow's 7th exception to his theory is well provided by the philosophical idea that the need to pursue the quest for ascendant quality is more fundamental than any basic need. In a similar analysis, it can be shown that this philosophical idea may also explain the remaining 6 exceptions to the Hierarchy of Needs theory which Dr. Maslow identified.

THE BATTLEFRONT OF NEEDS

The influence of needs occurs along a simultaneously shifting battlefront, spanning all of Maslow's Basic Needs, with a center of gravity and with decisive points corresponding to each basic need.

The Battlefront of Needs is the second tenet of the Ascendant Theory of Motivation. In The Battlefront of Needs we can see that the need to pursue the quest for ascendant quality is probably the incentive that causes progressive motivation within, in the direction of ascendant quality, and upwards, through

Maslow's hierarchy of basic needs. From this perspective, motivation across all needs is orchestrated as part of an interrelated, opportunistic strategy to optimize attainment of incentives and to maximize the satisfaction that doing so brings as various opportunities arise. This interpretation of Dr. Maslow's experimental observation is perhaps a more accurate representation of the true rationale for a person's progression through and upward through Maslow's Hierarchy of Needs. Dr. Maslow's theory explained this upward progression as follows:

> "Human needs arrange themselves in hierarchies of pre-potency. That is to say, the appearance of one need usually rests on the prior satisfaction of another, more pre-potent need. Man is a perpetually wanting animal."[167]

While Maslow's theory of the progression from one need to the next higher need appears plausible, the explanation that this happens because "man is a perpetually wanting animal" could perhaps be better explained by the proposition that there is an underlying need to pursue the quest for ascendant quality. This proposition doesn't necessarily contradict Maslow's explanation – it just adds an incentive of a quest for ascendant quality as an object of motivation. In the absence of this fundamental need explaining the progressiveness of need driven behavior there is no reason why a person wouldn't simply try to gratify the same need again or satisfy a lower level need all over again. A perpetually wanting animal shouldn't have a problem doing that. The point is, man is not an animal without an ascendant purpose and, as stated in the preamble to this chapter, no form of life is. The innovative progress of mankind in scientific endeavors and in quality of life over the centuries is ample evidence of the validity of this theory.

---

[167] Abraham Maslow, A Theory of Human Motivation, July 1943, Psychological Review, Vol. 50, 370-396

The third tenet of the Ascendant Theory of Motivation is that people develop personal strategies to accomplish satisfaction of their needs. Dr. Maslow's research appears to confirm the idea that strategy, acting as a center of gravity for purposeful behavior, is used in pursuit of needs (or goals).

> "... the most prepotent goal will monopolize consciousness and will tend of itself to organize the recruitment of the various capacities of the organism."[168]

Abraham Maslow

For example, if a man has a physiological need – hunger for example – he may obsess over the hunger and apply his best skills to forage for food, he may set a trap to catch food, he may barter for, beg for, borrow, or steal food, etc. Similarly, and with similar virtue or inelegance, other needs may be satisfied. All of these behaviors result from a strategic plan that he devises based on circumstances and resources, what he learns in the process, and the timeframe he has to find something. There's no way, other than through random luck, to satisfy a need without a strategy to do it. For this reason, we are significantly better off when we use our knowledge of strategy to improve our need gratification as well as to attain gratification from our creative effort.

The concept of personal strategy raises an additional point; although a person may be focused on a particular need at any time, most of the time our behavior is also in the service of other needs. This Battlefront of Needs can be

---

[168] Abraham Maslow, A Theory of Human Motivation, July 1943, Psychological Review, Vol. 50, 370-396

visualized in the same way that a Battlefront has been pictured on a Saile Chart. This underscores the point that where there are one or more needs a person will develop a personally optimal strategy to serve them as defined by their Battlefront of needs. Support for this proposition can be found in one of Dr. Maslow's observations:

> "Any motivated behavior, either preparatory or consummatory, must be understood to be a channel through which many basic needs may be simultaneously expressed or satisfied."[169]
>
> Abraham Maslow

In personal strategy, a need is satisfied when the individual's capabilities cannot provide the opportunity of further satisfaction or the individuals' desire to put forward additional effort in pursuit of a need is dissipated due to the prospect of achieving gratification more easily from pursuit of alternative needs. Need switching, like strategy switching, also occurs when personal strategy indicates that effort on the alternate need or needs may be more productive in pursuit of ascendant quality. As a person experiences life, and as attainment of incentives and satisfaction occur, needs grow and fade and the cycle of effort in pursuit of ascendant quality continues to evolve. The learning driven level of growth a person desires or aspires to in each Theme of Performance may be called the person's personal performance quality excellence definition. In this construct, a person will seek to reduce the risk of poor performance through strategies he or she can formulate and execute based on capabilities and resources available or attainable. This pursuit is their motivated behavior.

The realization that a strategy is at work in the process of satisfying needs changes the game in our approach to understanding motivated behavior. All the expertise we have developed in our study of the definition and structure of strategy, throughout this book, can be used as the basis for understanding the strategy exercised through motivated behavior and in the shaping of strategic communication to attain Coascendancy.

---

[169] Abraham Maslow, A Theory of Human Motivation, July 1943, Psychological Review, Vol. 50, 370-396

As our discussion has shown, the fundamental need to pursue the quest for ascendant quality provides the basis for personal strategy. By understanding personal strategy, the behaviors of people, organizations, and even nations, may be understood – at least with a degree of accuracy that's significantly better than speculation.

Implications:

- Satisfaction is the result of successfully performing behavior in accordance with personal strategy motivated by the need to pursue the quest for ascendant quality.

- Dissatisfaction is the result of the thwarting of the ability to successfully perform behavior in accordance with personal strategy.

  People will be content in satisfaction or will tolerate dissatisfaction until they devise a transformational personal strategy that's more satisfying and that's consistent with their approach to attainment of an improved quality of life.

- Motivation is caused by the opportunity to perform behavior in accordance with personal strategy in pursuit of the quest for ascendant quality.

  People are experts at spontaneously developing strategy. For most people, developing strategy resembles an instinctive talent. They pick the strategy which fits their resources and capabilities and which reduces their risk of poor performance against their personal concept of performance quality excellence -which is most consistent with their values. Personal values are weighed in the selection of a personal strategy and can cause people to favor or accelerate acceptance of a strategy or can cause them to disfavor or reject it. From among acceptable strategies, the one best strategy is selected based on expected performance in each Theme of Performance. It is towards successful performance of this one best strategy that a person's thoughts and behavior will be motivated.

- People adapt to the strategy that an organization projects to be suitable to its needs based on the quality of the strategic communication they receive and its coincidence with their personal strategy. When two or more parties, for example an employee, employer, and customer, perform this mutually beneficial negotiation and adaptation, this accomplishment should be recognized as Coascendancy.

- Productivity is how much a person produces in a certain amount of time. Productivity is optimized by the motivated behavior of an employee in pursuit of an organizational strategy for achieving ascendant quality.

- For creative effort, performance quality excellence, including productivity, to the best of a person's capabilities, can be achieved effectively and efficiently through Coascendancy.

- Performance Quality Excellence, in a Coascendant relationship, is satisfying to everyone.

- The process of evolution is a form of enrichment of a species which is symptomatic of the quest for ascendant quality applied to the mating or reproductive strategies of a living being.

Satisfaction in the workplace

A prominent study of satisfaction and dissatisfaction in the workplace was performed by Professor Frederick Herzberg of Case Western Reserve University. Professor Herzberg's Two Factor Theory was summarized in a popular article for the Harvard Business Review, in 1968, titled "One More Time: How do you Motivate Employees." In Professor Herzberg's study, participants "were asked what job events had occurred in their work that had led to extreme satisfaction or extreme dissatisfaction on their part." Dr. Herzberg's findings are charted in figure 7-2.

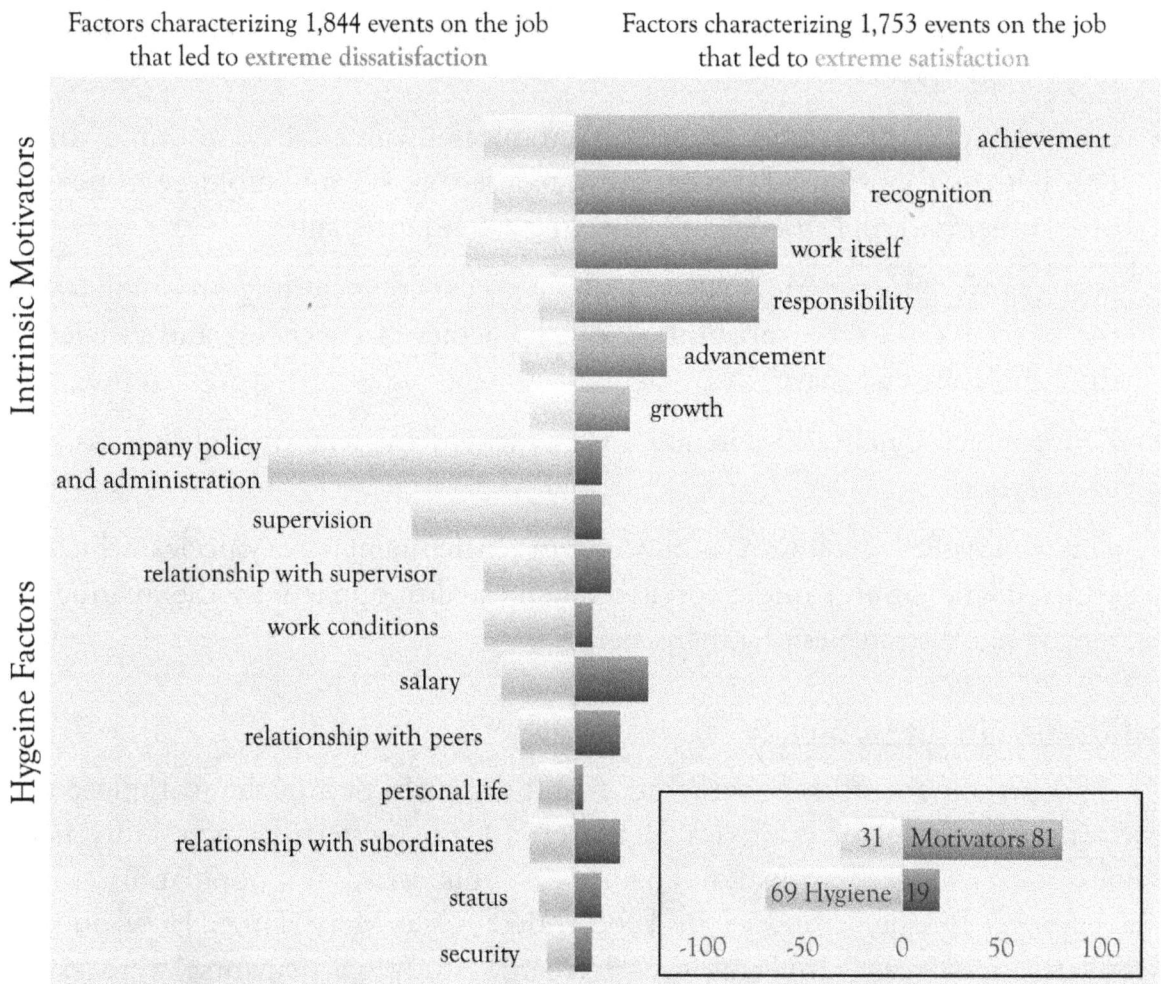

Figure 7-2; Frederick Herzberg's classification of events leading to job satisfaction or job dissatisfaction.

If we were to confirm the proposition that the need to pursue the quest for ascendant quality is a fundamental need, considering data from Professor Herzberg's study, we would expect to validate the following 3 hypotheses:

1. There is significant satisfaction in events that serve to sustain, enable, and advance personal strategy.
2. The need most often associated with the feeling of satisfaction is the need to be on the quest for ascendant quality.
3. There is significant dissatisfaction in events that obstruct or create a risk of poor performance in a person's ability to sustain, enable or advance their personal strategy.

Professor Herzberg's study indicates that a high percentage of events leading to extreme satisfaction were found in the categories of Achievement, Recognition, Work Itself, Responsibility, Advancement and [Possibility of] Growth. With interpretation, these results validate our thesis that there is an underlying, fundamental, need to pursue the quest for ascendant quality and that doing so brings satisfaction.

- Achievement, the highest category of satisfaction, is consistent with achieving progress in the quest for ascendant quality.

- Recognition is an event that consolidates the feeling of progress in the pursuit of ascendant quality and socializes this fact helping the individual benchmark progress in ascendancy.

- Work itself is satisfying because it's what the person does to achieve ascendancy.

- Responsibility is satisfying because it encapsulates the range of things the person is capable of doing and is free to do that can bring satisfaction and ascendancy – the more responsibility the better as long as it's within the person's capabilities.

- Advancement is similar to responsibility and recognition because the authority to implement activities to achieve ascendancy is recognizably increased with advancement and in some circumstances the range of opportunities to perform ascendant activities may be extended.

- The possibility of growth provides hope for advancement in the future. The Possibility of Growth is close in characterization to what is meant by the quest for ascendant quality. It's a lifelong need and not one that would give rise to "events on the job providing extreme satisfaction." Rather it's a sublime, long term satisfaction like pride and contentment. Herzberg was aware of this long term effect when he wrote:

> "In contrast to recognition and achievement, where long-range and short-range attitude changes are of about equal frequency, the factor of possible growth shows more than a two-to-one ratio in favor of the long-range positive changes in feeling toward a job. The great frequency with which this need was verbalized indicates that the feeling that personal or professional growth was possible bulked

very large in the psychological reactions of the individual to the kinds of situations he described as central to a high."[170]

Dr. Frederick Herzberg
1959

In this set of events leading to extreme satisfaction – which Herzberg called "intrinsic motivators" – it appears that they are all events that serve to sustain, enable, and advance personal strategy – which is what we expected to find. Professor Herzberg also found that:

> "Apparently, the feeling of growth in stature and responsibility is still the most exciting thing that can happen to someone in our society."[171]

Dr. Frederick Herzberg
1959

This finding is a clear indication that the fundamental need to be on the quest for Ascendant Quality is at the heart of events leading to extreme satisfaction on the job.

We can also assess the set of events leading to extreme dissatisfaction, in a similar analysis, as follows:

- <u>Company policy and administration</u> achieve a lot of dissatisfaction because they impose rules. When the rules slow progress, distract the sense of purpose, or are inconsistent with worker values or their understanding of the proper way to accomplish a task, they are dissatisfying. When these rules help to improve productivity or remove chaos they are probably quite satisfying.

- <u>Supervision</u> is dissatisfying to a subordinate when the subordinate is derailed in their personal quest for ascendancy by requirements imposed by the

---

[170] Frederick Herzberg, Bernard Mausner, Barbara Bloch Snyderman, The Motivation to Work, New York : London : Wiley ; Chapman & Hall, 1959-2010, p. 68.
[171] Frederick Herzberg, Bernard Mausner, Barbara Bloch Snyderman, The Motivation to Work, New York : London : Wiley ; Chapman & Hall, 1959-2010, p. 130.

supervisor that curtail their freedom. When an employee is supported and enabled to thrive, supervision may be viewed with great satisfaction.

- The <u>relationship with supervisor</u> is dissatisfying when the supervisor exploits subordinates by driving them to do unnecessary or undesirable work or by stealing ideas for their own gain thereby preventing the ascendant feeling and satisfaction of subordinates. Supervisors can support employee satisfaction by enabling ascendancy through training, assignments to challenging tasks, collaboration, and joint learning, etc.

- <u>Work conditions</u> are dissatisfying when they introduce hindrances to performance or are unhealthy, and are probably satisfying when they have the capability to facilitate workflow of high quality and support productive interpersonal interactions and communication.

- <u>Salary</u> is almost always motivating – but not necessarily satisfying on the job to the same extent as it's satisfying off the job, at home, where the salary is actually needed. As you may discern for yourself, satisfaction is the aftereffect of the sequence of being motivated, planning to do something, doing something, and then feeling satisfied. The better, or higher, a salary is, the better the employee's potential for enjoying a lifestyle that is more comfortable and supportive of ascendancy. The lower, or worse, a salary is, the more likely an employee is to be dissatisfied and quit the job for a better paying one – even if the employee feels a sense of achievement, gets recognition, feels the work is challenging, etc.

While studying people to determine what satisfies them is easier than making inferences based on animal studies – because people can answer questions directly – people don't always provide reliable answers. The question of salary is one such case where interviewees may give misleading answers – especially if by saying that they are satisfied with their salary they may end up losing future raises. In this sense, asking employees if they're satisfied with their salary is like asking prison inmates if they're guilty.

An example you've seen for yourself of money-driven motivation is that in many doctor's offices nowadays you may find a doctor treating 10 patients per hour. The Doctor's probably not doing this because he or she wants to bring good health to the masses – at least not as much as he or she is motivated to earn money to repay student loans, pay malpractice insurance, or pay for a nice vacation.

People will also report dissatisfaction when, relative to people in other professions requiring similar education and experience, they feel underpaid. While I can understand the affection business leaders may have for the idea that they can motivate their employees without paying them more or working with them to pursue each other's best destiny, the research tells a different story.

- Relationship with peers, personal life, relationship with subordinates, status, and security can all be satisfying, unsatisfying or dissatisfying depending on their relationship to personal strategy. The fact that personal life scored so poorly in count of events on the job that led to extreme satisfaction may have a lot to do with the workplace not being the place where the personal life of the employee takes place. We may find many of these employees quite satisfied at home.

Herzberg's study wasn't intended to test whether our hypothesized need to be on the quest for ascendant quality is valid; however, analysis of his study data provides a good amount of support for this hypothesis. As with all new theories, new experiments are needed to improve our understanding of the need to pursue the quest for ascendant quality.

Productivity

One of the most significant questions businesses need to answer, in order to produce competitive and successful products and services, is – "How do you make employees more productive?

The answer is not necessarily to redesign the production system or even the incentive system to achieve employee satisfaction, the answer must be found in doing what actually causes an increase in productivity – the rate of throughput of products and services of high quality. If redesigning the productive system or incentive system is what it takes then, by all means, improve them to assure higher productivity and profit. If employee satisfaction is also achieved in the process then you at least know which improvements are well worth retaining.

258

Using the Ascendant Theory of Motivation developed so far, it is clear that the answer to the question "How do you make employees more productive?" must build on the premise that employees have personal strategies and that the underlying need of employees personal strategies is ascendant quality. If the organization can also be said to hold these values, the question should be answered by simply saying "employees are more productive in their work effort when productivity in their work effort coincides with pursuit of their personal strategy on the quest for ascendant quality." Considering this model in an operational context, we should expect to see increases in productivity caused for the duration, and with the accompanying employee satisfaction, at about the rate depicted in figure 7-3.

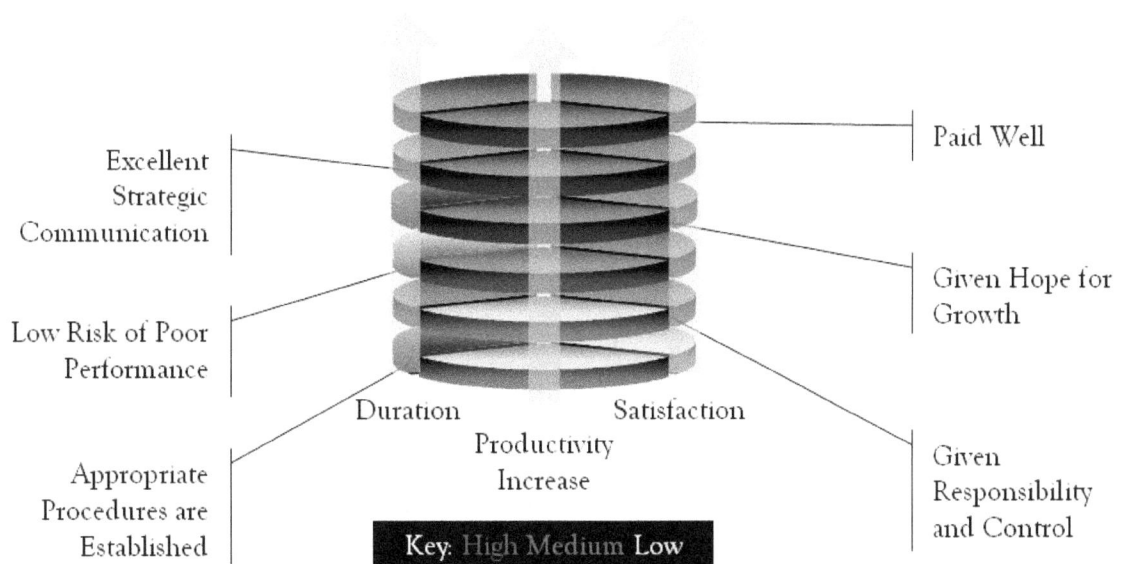

Figure 7-3, Factors causing increased productivity and the duration of their coincident effect on employee satisfaction.

Innovation

The cause of innovation is the coincidence of the realization that ascendant quality in the challenge of interest lies beyond the bounds of the state of the art, as the state of the art is currently known, and your feeling of the influence of an incentive propelling you to apply your talent and resources to contrive of a strategy or solution that lies beyond the state of the art.

Success in innovation takes persistent effort and the confidence in success that comes from deep understanding of the multiple, simultaneous, personal, organizational, and customer strategies which the solution must advance. For this reason, through collaboration in an organization, we can add improved innovative capability to employee satisfaction, improved productivity and value as benefits of Coascendancy.

The following excerpt of a letter from Vincent van Gogh to his brother Theo, reflecting on some things he has learned about painting natural scenes, is very enlightening about innovation, the drive to create, and the cognitive experience combined with the actions and talents contributed by an artist to his creative effort.

"...While painting it I said to myself: I must not go away before there is something of an autumn evening air about it, something mysterious, something serious.

But as this effect does not stay, I needed to paint quickly - the figures were painted in at once

Figure 7-4;
Vincent van Gogh:
Portrait of the Artist,
1889

with a few strong strokes with a firm brush. It struck me how firmly those little stems were rooted in the ground. I began on them with a brush, but because the base was already so clotted, a brush-stroke was lost in it - so I squeezed the roots and trunks in from the tube, and modeled it a little with the brush.

Yes - now they stand there rising from the ground, strongly rooted in it. In a certain way I am glad I have not learned painting, because then I might have learned to pass by such effects as this. Now I say, no, this is just what I want, if it is impossible, it is impossible; I will try it, though I do not know how it should be done. How I paint it I do not know myself. I sit down with a white board before the spot that strikes me, I look at what is before me, I say to myself that that white board must become something; I come back dissatisfied – I put it away, and when I have rested a little, I go to look at it with a kind of fear. Then I am still dissatisfied, because I still have too clearly in my mind that splendid subject, to be satisfied with what I made of it. But after all I find in my work an echo of what struck me. I see that nature has told me something, has spoken to me, and that I have to put it down in shorthand. In my shorthand there may be words that cannot be deciphered, there may be mistakes or gaps, but there is something in it of what wood or shore or figure has told me, and it is not a tame or conventional language, proceeding less from nature itself than from a studied manner or a system..."

Vincent Van Gogh
Letter to his brother Theo, 1882

The Creation of Value

"Now I think, speaking roughly, by leadership we mean the art of getting someone else to do something that you want done because he wants to do it, not because your position of power can compel him to do it, or your position of authority."[172]

Dwight D. Eisenhower
1954

If we believe that people have a need to pursue the quest for ascendant quality, they must be executing a personal strategy with the intention of meeting all of their needs. The organization is a mechanism to gratify some of these needs and the home and life outside the office gratifies others. In advancing the quality of an organization, leaders should be interested in a policy towards employees that serves to encourage and promote the common interests of personal satisfaction, aspiration, innovation, motivation, productivity, long term employment, competitive wages, and quality of life in the workplace and beyond. In this way, excellent Coascendancy can be achieved for the benefit of employees,

---

[172] Dwight D. Eisenhower: "Remarks at the Annual Conference of the Society for Personnel Administration," May 12, 1954. Online by Gerhard Peters and John T. Woolley, The American Presidency Project. http://www.presidency.ucsb.edu/ws/?pid=9884.

the organization and the organizations' customers in accordance with their shared vision of performance quality excellence.

The Coascendancy Totem, pictured in figure 7-5, represents an operational view of our socially accepted integration of organizational, employee, and customer strategies in a commercial venture – along with the wedges obstructing motivated and efficient performance.

Beginning at the lower left, employees' personal goals, organizational goals, and customer goals and expectations are shown incentivizing an organizational effort. Strategic communication is used to assist leadership and organization of the productive effort. Where the organizational competency in strategic communication is poor or confusing it dilutes the influence of the incentive and serves as an obstruction to the efficiency and effectiveness of the productive effort. This is represented in the Totem as a pink wedge between the incentive and the ability to formulate a high quality motive with regard to the incentive. Poor organizational planning for the productive effort is another competency lapse – shown as a pink wedge – obstructing or frustrating employees

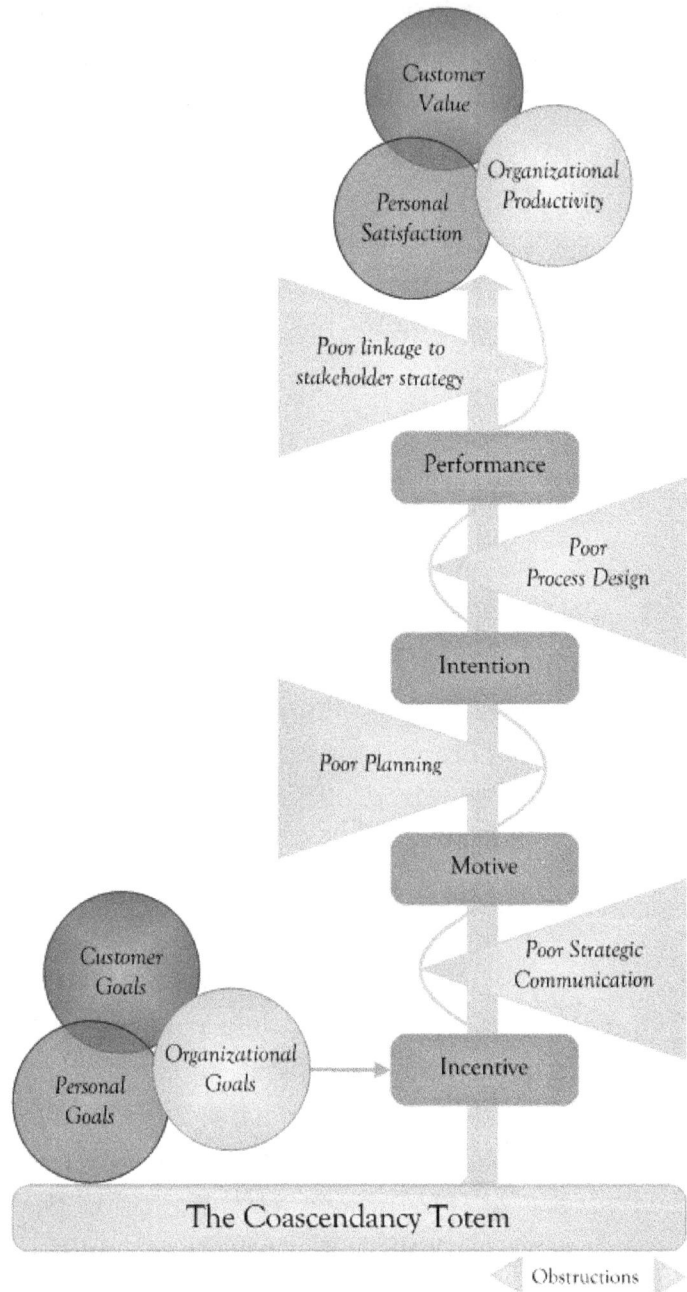

Figure 7-5; The Coascendancy Totem.

262

from organizing, planning and setting intentions for their effort. (Elements of excellent strategic communication were discussed in chapters 1 – 3.) Poor process design is an additional obstruction to performance quality excellence. When employees are not able to change poor process design they may become depressed under a psychological theory called Learned Helplessness that was developed by Psychologist Martin Seligman. Learned helplessness may have an enduring negative effect that prevents employees from correcting problems even after they are enabled to change process design. In this circumstance, process assessment using a Saile chart can help focus employees on necessary improvements.

The rewards to the stakeholders – the employees in terms of personal satisfaction, the organization in terms of productivity, and the customer(s) in terms of value – depend on the results of all the activities in the Coascendancy Totem and how well goals are met throughout the productive effort and how well the goals represent the strategies of the parties they represent. A key observation in the Coascendancy Totem is that, by removing the obstructions, everyone gains what they want – a satisfying place to work, a productive workplace, and products and services that provide value to customers. An additional consequence of successful strategic alignment is the ability to transform, in depth, through quality improvement and transformational innovation, because of a depth of mutual understanding and coincidence of needs. Attainment of this condition is dependent on countless environmental and interpersonal factors, however, in an organizational setting, it is the center of gravity of the Coascendant strategy for which the organization, employees, and customers strive which has the potential to harmonize creative effort and bring this ideal condition to each of them.

In conditions of quality excellence, in the absence of the risk of poor performance, where the stage is set for Ascendant Quality to emerge, individuals are probably at their best state of motivation and are probably in their happiest and most gratifying, normal, condition because the need to be on the quest for ascendant quality is a fundamental need which this condition fulfills.

Chapter 6:

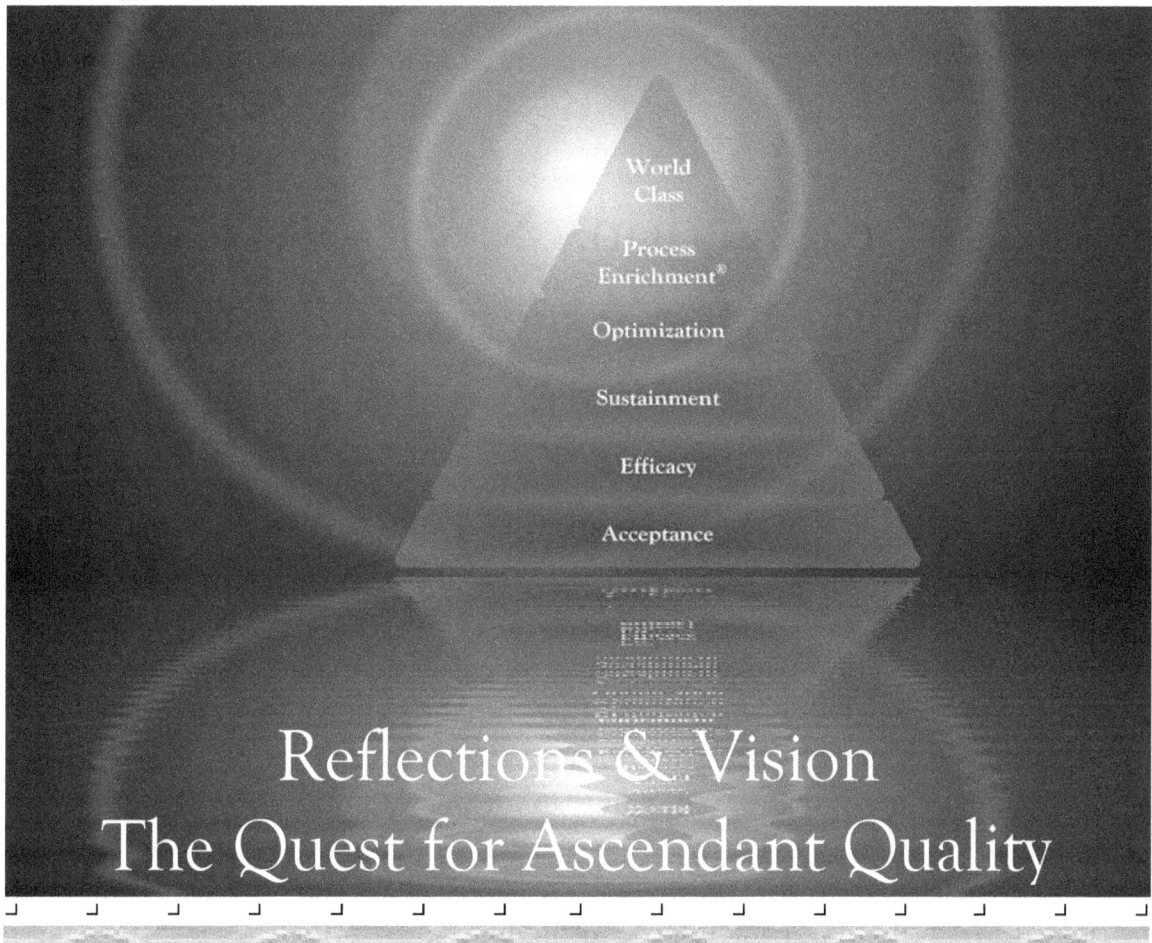

Reflections & Vision
The Quest for Ascendant Quality

World Class
Process Enrichment®
Optimization
Sustainment
Efficacy
Acceptance

"Tell us a good action story!"
Oliver Elias
2001 (Age 4)

Chapter 6:

Reflections & Vision
The Quest for Ascendant Quality

PERSPECTIVES

▶ The evident effect for which civilization demands ascendant quality
▶ The divergence between the Process Enrichment philosophy and the "Absolutes" of quality management in the 20th century
▶ A working vision of the role of quality in ascendant cultures for the foreseeable future
▶ A "good action story" for Oliver and Victoria and for those of us, throughout the world, on the quest for ascendant quality

The ever evolving ability of civilizations to provide a better quality of life everywhere and for everyone is the evident effect for which humanity has enjoyed, and will continue to enjoy, the quest for ascendant quality. Our common and fundamental human need, to be on the quest for ascendant quality, ensures that wherever in the world ascendant cultures exist we will continue to be energized and inspired to integrate new and improved technologies in our strategies, to actualize our most far reaching creative ambitions, and to gainfully resolve our most daunting challenges.

In changing the way we recognize the fundamental concepts which shape our ever evolving and ever ascending ideal of quality excellence and in defining the value system central to our interactions in commerce we have drawn ourselves a long way from the ideas and philosophy of the last century which were once considered sacrosanct "absolutes." Philip Crosby's "four absolutes of quality management," listed in table 6-1, which energized the quality revolution of the late 20th century, are a significant milestone to reflect upon to appreciate the extent of our journey in the early 21st century with the dawning of today's Process Enrichment philosophy.

As dramatic experiences of euphoric success and of weighty failure throughout history have shown and as an exploration and renewed appreciation of our own natural talents in formulating strategy has shown, what we have long embraced, intuitively, as an integrated conceptual system is reflected and expressed consistently in the Process Enrichment philosophy. Our study has shown that the development and verification of each new proposed element of our conceptual system has served to build another and another – as must be expected with significant new conceptual ideas. In consequence, this progression has amounted to a new framework with which a holistic and useful understanding of the elements of quality can be readily imagined.

Absolutes of Quality Management in the 20th Century		Quality's Conceptual System in the 21st Century and Beyond
The First Absolute	The Definition of Quality is Conformance to Requirements, Not Goodness	The definition of quality is "The ability of performance to enact a strategy."
The Second Absolute	The System for Causing Quality is Prevention, Not Appraisal	The system for causing quality is an integrated conceptual system in which competencies and resources are organized for the purpose of attaining the incentives of a customer's valued strategy at a minimum of risk.
The Third Absolute	The Performance Standard is Zero Defects, Not "Close Enough"	The performance standard is quality excellence – which represents the ability of performance to enact a customer's valued strategy, absent risks of poor performance, at a price customers are ready, willing, and able to pay and that is of a desirable profit (or other incentive) for the producer of the creative effort.
The Fourth Absolute	The Measurement of Quality is the Price of Nonconformance, Not Indexes	The measurement of quality is the absence of the risk of poor performance – which is achieved when there is a minimum of deviation from quality excellence.

Table 6-1; Divergences between the Process Enrichment philosophy and Philip Crosby's four absolutes of quality management.[173]

Having now dialed in all the tumblers to their correct position, on the enigmatic vault containing our philosophy, and having opened the vault, we find ourselves in the presence of a great discovery. It is a discovery of an interrelated system which stands alone as a direct approach to creative innovation and quality improvement. For me, as I hope it has been for you, this culmination of ideas, strange at first and second nature at last, revealing the Process Enrichment philosophy, is a personal validation of the paradox expressed by the famous author Arthur Koestler.

---

[173] Crosby, Philip B. *Quality Without Tears: The Art of Hassle-Free Management*. New York: McGraw-Hill, 1984.

"...the more original a discovery the more obvious it seems afterwards. The creative act is not an act of creation in the sense of the Old Testament. It does not create something out of nothing; it uncovers, selects, re-shuffles, combines, synthesizes already existing facts, faculties, skills. The more familiar the parts, the more striking the new whole."[174]

Arthur Koestler
1964

As this phase of our journey on an adventure of discovery comes to a close, I will sum up a working vision of quality that may serve long into the foreseeable future: *When creative effort is pursued with the beauty of quality, the enlightenment of risk, and the gravitational attraction of a valued strategy in mind, and when leaders act to cultivate Coascendancy among those who gainfully pursue the quest for ascendant quality together, quality can improve well beyond its previous bounds and civilization may hasten its ascent to the great heights of its destiny.*

While the new philosophy and methods presented throughout this book bear a transformative message in advancement of ascendant quality, it's we the people who join the quest for ascendant quality, in our own way, that have the power to carry forth the transformation and achieve what was once impossible. In the long view of history these are our times and our quest for ascendant quality is our good action story.

---

[174] Koestler, Arthur. 1964. The Act of Creation, p.120. Penguin Books, New York.

www.ingramcontent.com/pod-product-compliance
Lightning Source LLC
Chambersburg PA
CBHW061350210326
41598CB00035B/5944